CONCEPTS AND CHALLENGES

ENERGY

Leonard Bernstein ◆ Martin Schachter ◆ Alan Winkler ◆ Stanley Wolfe

Stanley Wolfe
Project Coordinator

GLOBE FEARON
Pearson Learning Group

The following people have contributed to the development of this product:

Art and Design: Evelyn Bauer, Susan Brorein, Tracey Gerber, Bernadette Hruby, Carol Marie Kiernan, Mindy Klarman, Judy Mahoney, Karen Mancinelli, Elbaliz Mendez, April Okano, Dan Thomas, Jennifer Visco

Editorial: Stephanie P. Cahill, Gina Dalessio, Nija Dixon, Martha Feehan, Theresa McCarthy, Maurice Sabean, Marilyn Sarch, Maury Solomon, Jeffrey Wickersty, Shirley C. White, S. Adrienn Vegh-Soti

Editorial Services: Thomas Daning, Richetta Lobban

Manufacturing: Mark Cirillo, Tom Dunne

Marketing: Douglas Falk, Maureen Christensen

Production: Irene Belinsky, Linda Bierniak, Carlos Blas, Karen Edmonds, Cheryl Golding, Leslie Greenberg, Roxanne Knoll, Susan Levine, Cynthia Lynch, Jennifer Murphy, Lisa Svoronos, Susan Tamm

Publishing Operations: Carolyn Coyle

Technology: Jessie Lin, Joanne Saito, Ellen Strain

About the Cover: Energy is needed to make things happen. The light coming from the optical fibers shown in the larger photograph is produced by very tiny packets of energy called photons. The energy of the water wave shown in the smaller photograph is sent across entire oceans by moving particles of water.

ISBN: 0-13-024197-0

Printed in the United States of America

1 2 3 4 5 6 7 8 9 10 06 05 04 03

Globe Fearon
Pearson Learning Group

1-800-321-3106
www.pearsonlearning.com

Acknowledgments

Science Consultants

Jonathan Cohen
Science Teacher
Longfellow Arts and Technology
Magnet Middle School
Berkeley, California

Kenneth S. Fink
Liberty Science Center
Jersey City, New Jersey

Laboratory Consultants

Sean Devine
Science Teacher
Ridge High School
Basking Ridge, New Jersey

Vincent Dionisio
Science Teacher
Clifton High School
Clifton, New Jersey

Gregory Vogt, Ph.D.
Associate Professor, Civil Engineering
Colorado State University
Fort Collins, Colorado

Reading Consultant

Sharon Cook
Consultant
Leadership in Literacy

Internet Consultant

Janet M. Gaudino
Science Teacher
Montgomery Middle School
Skillman, New Jersey

ESL/ELL Consultant

Elizabeth Jimenez
Consultant
Pomona, California

Content Reviewers

Sharon Danielsen (pp. 36, 37)
Site Manager
Darrin Fresh Water Institute
Rensselaer Polytechnic Institute
Troy, New York

T. Ferbel (Ch. 5)
Professor of Physics
University of Rochester
Rochester, New York

Ernest Freund (Chs. 2–4)
Lecturer
University of California at Santa Barbara
Santa Barbara, California

Dr. Raymond C. Turner (Ch. 6)
Alumni Distinguished Professor Emeritus of Physics
Department of Physics and Astronomy
Clemson University
Clemson, South Carolina

Todd Woerner (Ch. 1)
Department of Chemistry
Duke University
Durham, North Carolina

Teacher Reviewers

Leonard GeRue
Hanshaw Middle School
Modesto, California

Charles Sehulster
Science Teacher
Horace Greeley High School
Chappaqua, New York

Contents

Scientific Skills and Investigations Handbooks

Chapter 1 Heat 15

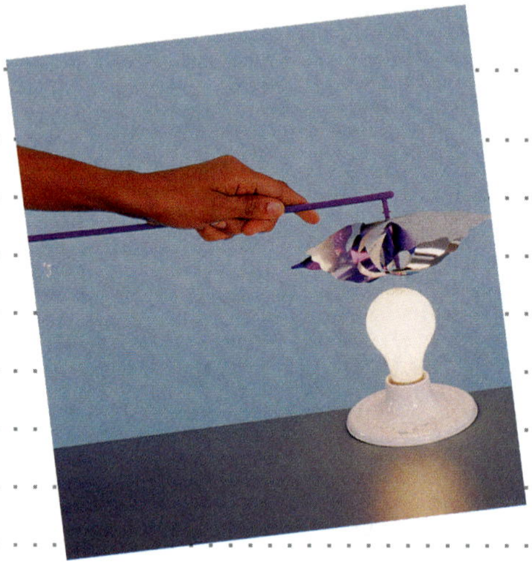

Chapter 2 Waves 41

Chapter 3 Sound 59

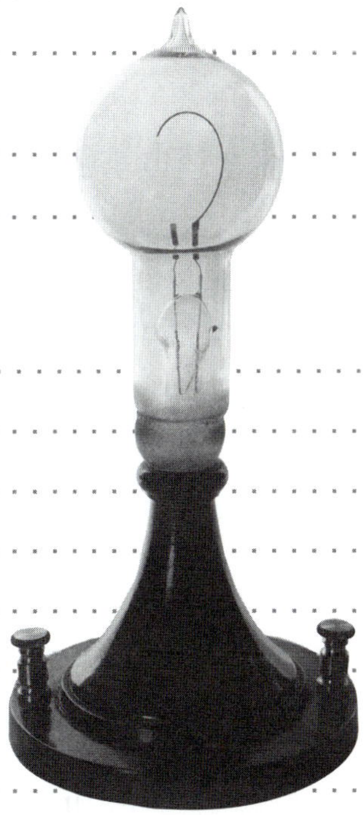

Features

Hands-On Activities

How Do They Know That?

Integrating the Sciences

Real-Life Science

People in Science

Science and Technology

INVESTIGATE

Web InfoSearch

What are scientific skills?

People are naturally curious. They want to understand the world around them. The field of science would probably not exist if it were not for human curiosity about the natural world.

People also want to be able to make good guesses about the future. They want to know how to use alternative forms of energy. They want to improve technology and communications.

Scientists use many skills to explore the world and gather information about it. These skills are called science process skills. Another name for them is science inquiry skills.

Science process skills allow you to think like a scientist. They help you identify problems and answer questions. Sometimes they help you solve problems. More often, they provide some possible answers and lead to more questions. In this book, you will use a variety of science process skills to understand the facts and theories in physical science.

Science process skills are not only used in science. You compare prices when you shop and you observe what happens to foods when you cook them. You predict what the weather will be by looking at the sky. In fact, science process skills are really everyday life skills that have been adapted for problem solving in science.

1 NAME: What is the name for the skills scientists use to solve problems?

▲ **Figure 1** Scientists use science process skills to understand what makes a nuclear power plant run safely, how robots work in spaces too small for humans, and why communications are better using fiber optics.

Contents

1 Observing and Comparing

Making Observations An important part of solving any problem is observing, or using your senses to find out what is going on around you. The five senses are sight, hearing, touch, smell, and taste. When you look at the properties of an ore or watch an ice cube melt, you are observing. When you observe, you pay close attention to everything that happens around you.

Scientists observe the world in ways that other scientists can repeat. This is a goal of scientific observation. It is expected that when a scientist has made an observation, other people will be able to make the same observation.

2 **LIST:** What are the five senses?

Comparing and Contrasting Part of observing is comparing and contrasting. When you compare data, you observe the characteristics of several things or events to see how they are alike. When you contrast data, you look for ways that similar things are different from one another.

▲ **Figure 2** Silver and aluminum are alike in many ways. They also have many differences.

3 **COMPARE/CONTRAST:** How are a bar of aluminum and a bar of silver alike? How are they different?

Using Tools to Observe Sometimes an object is too small or too distant to see with your eyes alone. Often, special tools are needed for making observations. Sometimes scientists use tools to make observations of things like radio waves or X-rays that are outside the range of our senses. Telescopes, spectrometers, microscopes, and magnifying glasses are all examples of tools that help with scientific observations.

▲ **Figure 3** Examining a slide with a magnifying glass

4 **INFER:** What are some things that scientists might need a microscope to see?

Hands-On Activity

MAKING OBSERVATIONS

You and a partner will need 2 shoeboxes with lids, 2 rubber bands, and several small objects.

1. Place several small objects into the shoebox. Do not let your partner see what you put into the shoebox.
2. Cover the shoebox with the lid. Put a rubber band around the shoebox to keep the lid on.
3. Exchange shoeboxes with your partner.
4. Gently shake, turn, and rattle the shoebox.
5. Try to describe what is in the shoebox without opening it. Write your descriptions on a sheet of paper.

Practicing Your Skills

6. **IDENTIFY:** What science process skill did you use?
7. **IDENTIFY:** Which of your senses was most important to you?
8. **ANALYZE:** Direct observation is seeing something with your eyes or hearing it with your ears. Indirect observation involves using a model or past experience to make a guess about something. Which kind of observation did you use?

2 Classifying Data

Key Term

data: information you collect when you observe something

Collecting and Classifying Data The information you collect when you observe something is called **data.** The data from an experiment or from observations you have made are first recorded, or written down. Then, they are classified.

When you classify data, you group things together based on how they are alike. This information often comes from making comparisons as you observe. You may classify by size, shape, color, use, or any other important feature. Classifying data helps you recognize and understand the relationships between things. Classification makes studying large groups of things easier. For example, physical scientists use classification to organize different types of elements.

5 **EXPLAIN:** How can you classify data?

Hands-On Activity

ORGANIZING LIQUIDS

You will need 10 to 15 jars or bottles filled with a variety of liquids.

1. Carefully examine the liquids in the containers. Observe their color, thickness, and composition. Notice what happens when you gently shake the bottle. What happens when you let the liquid settle?
2. Make a system for classifying the liquids.
3. Categorize all the liquids.
4. Write a description of how you would use your classification system to classify a new liquid that you have never seen before.

Practicing Your Skills

5. **ANALYZE:** How did you classify the liquids?
6. **EXPLAIN:** Why is a classification system useful?

3 Modeling and Simulating

Key Terms

model: tool scientists use to represent an object or process

simulation: computer model that usually shows a process

Modeling Sometimes things are too small to see with your eyes alone. Other times, an object is too large to see. You may need a model to help you examine the object. A **model** is a good way to show what a very small or a very large object looks like. A model can have more details than what may be seen with just your eyes. It can be used to represent a process or an object that is hard to explain with words. A model can be a three-dimensional picture, a drawing, a computer image, or a diagram.

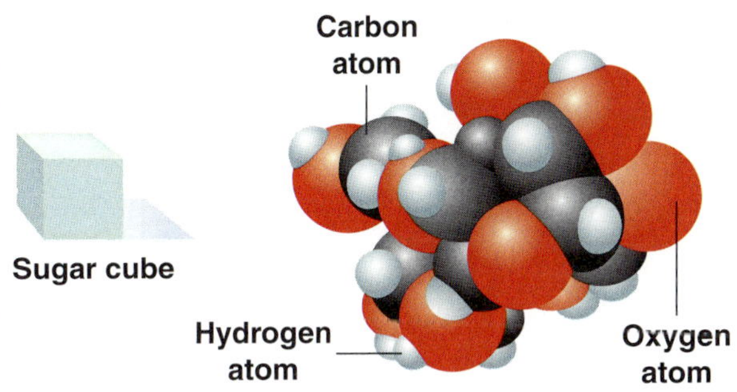

▲ **Figure 4** Some schools have molecule-building kits. Each atom is color-coded according to the element it represents.

6 **DEFINE:** What is a model?

Simulating A **simulation** is a kind of model that shows a process. It is often done using a computer. You can use a simulation to predict the outcome of an experiment. Scientists use simulations to study everything from chemical reactions to the global climate.

7 **DEFINE:** What is a simulation?

4 Measuring

Key Terms

unit: amount used to measure something

meter: basic unit of length or distance

mass: amount of matter in something

gram: basic unit of mass

volume: amount of space an object takes up

liter: basic unit of liquid volume

meniscus: curve at the surface of a liquid in a thin tube

temperature: measure of the amount of heat energy something contains

Two Systems of Measurement When you measure, you compare an unknown value with a known value using standard units. A **unit** is an amount used to measure something. The metric system is an international system of measurement. Examples of metric units are the gram, the kilometer, and the liter. In the United States, the English system and the metric system are both used. Examples of units in the English system are the pound, the foot, and the gallon.

There is also a more modern form of the metric system called SI. The letters *SI* stand for the French words *Système International.* Many of the units in the SI are the same as those in the metric system.

The metric and SI systems are both based on units of ten. This makes them easy to use. Each unit in these systems is ten times greater than the unit before it. To show a change in the size of a unit, you add a prefix to the unit. The prefix tells you whether the unit is larger or smaller. For example, a centimeter is ten times bigger than a millimeter.

PREFIXES AND THEIR MEANINGS	
kilo-	one thousand (1,000)
hecto-	one hundred (100)
deca-	ten (10)
deci-	one-tenth (1/10)
centi-	one-hundredth (1/100)
milli-	one-thousandth (1/1,000)

◀ **Figure 5**

8 IDENTIFY: What are two measurement systems?

Units of Length Length is the distance from one point to another. In the metric system, the basic unit of length or distance is the **meter.** A meter is about the length from a doorknob to the floor. Longer distances, such as the distances between cities, are measured in kilometers. A kilometer is 1,000 meters. Centimeters and millimeters measure shorter distances. A centimeter is 1/100 of a meter. A millimeter is 1/1,000 of a meter. Figure 6 compares common units of length. It also shows the abbreviation for each unit.

SI/METRIC UNITS OF LENGTH	
1,000 millimeters (mm)	1 meter (m)
100 centimeters (cm)	1 meter
10 decimeters (dm)	1 meter
10 millimeters	1 centimeter
1,000 meters	1 kilometer (km)

▲ **Figure 6**

Length can be measured with a meter stick. A meter stick is 1m long and is divided into 100 equal lengths by numbered lines. The distance between each of these lines is equal to 1 cm. Each centimeter is divided into ten equal parts. Each one of these parts is equal to 1 mm.

▲ **Figure 7** A meter stick is divided into centimeters and millimeters.

9 CALCULATE: How many centimeters are there in 3 m?

Measuring Area Do you know how people find the area of the floor of a room? They measure the length and the width of the room. Then, they multiply the two numbers. You can find the area of any rectangle by multiplying its length by its width. Area is expressed in square units, such as square meters (m^2) or square centimeters (cm^2).

$$\text{Area} = \text{length} \times \text{width}$$

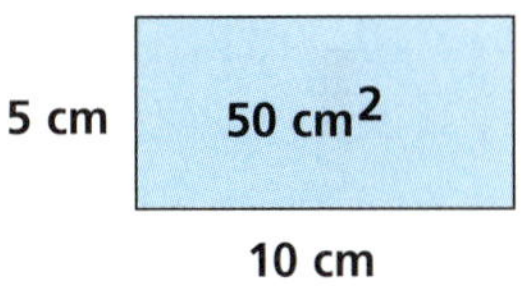

◀ **Figure 8** The area of a rectangle equals length times width.

10 CALCULATE: What is the area of a rectangle 12 cm by 6 cm?

Mass and Weight The amount of matter in something is its **mass**. The basic metric unit of mass is called a **gram** (g). A paper clip has about 1 g of mass. Mass is measured with an instrument called a balance. A balance works like a seesaw. It compares an unknown mass with a known mass.

One kind of balance that is commonly used to measure mass is a triple-beam balance. A triple-beam balance has a pan. The object being measured is placed on the pan. The balance also has three beams. Weights, called riders, are moved along each beam until the object on the pan is balanced. Each rider gives a reading in grams. The mass of the object is equal to the total readings of all three riders.

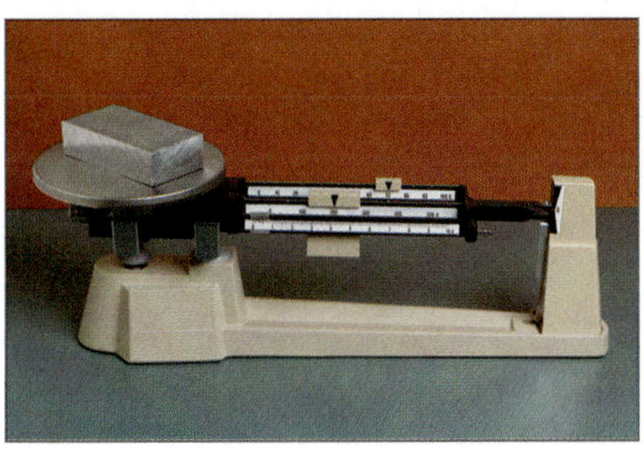

◀ **Figure 9** A triple-beam balance

Mass and weight are related; however, they are not the same. The weight of an object is a measure of Earth's pull of gravity between Earth and that object. Gravity is the force that pulls objects toward the center of Earth. The strength of the pull of gravity between two objects depends on the distance between the objects and how much mass they each contain. So, the weight of an object changes as its mass changes and as its distance from the center of Earth changes.

11 IDENTIFY: What instrument is used to measure mass?

Volume The amount of space an object takes up is its **volume.** You can measure the volume of liquids and solids. Liquid volume is usually measured in **liters.** Soft drinks in the United States often come in 2-liter bottles.

A graduated cylinder is used to measure liquid volume. Graduated cylinders are calibrated, or marked off, at regular intervals. Look at Figure 10. It shows a graduated cylinder. On this graduated cylinder, each small line is equal to 0.05 mL. The longer lines mark off every 0.25 mL up to 5.00 mL. However, every graduated cylinder is not marked in this manner. They come in different sizes up to 2,000 mL with different markings.

Always read the measurement at eye level. If you are using a glass graduated cylinder, you will need to read the mark on the graduated cylinder closest to the bottom of the meniscus. A **meniscus** is the curve at the surface of a liquid in a thin tube. A plastic graduated cylinder does not show a meniscus.

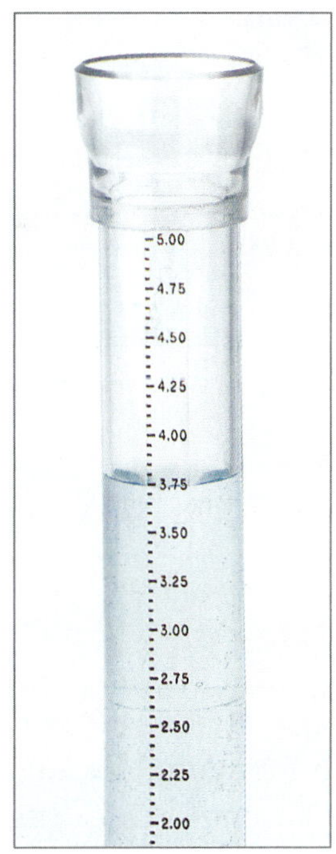

▲ **Figure 10** This glass graduated cylinder shows a meniscus.

The volume of solid objects is often measured in cubic centimeters. One cubic centimeter equals 1 mL.

Look at Figure 11. Each side of the cube is 1 cm long. The volume of the cube is 1 cubic centimeter (1 cm^3). Now, look at the drawing of the box in Figure 12. Its length is 3 cm. Its width is 2 cm. Its height is 2 cm. The volume of the box can be found by multiplying length by width by height. In this case, volume equals 3 × 2 × 2. Therefore, the volume of the box is 12 cm^3.

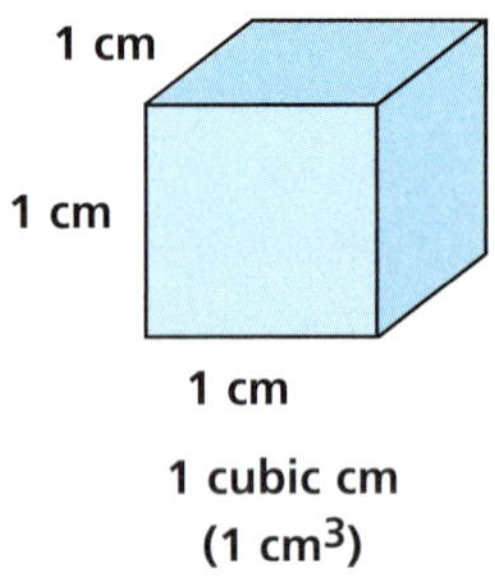

◀ **Figure 11** A cube

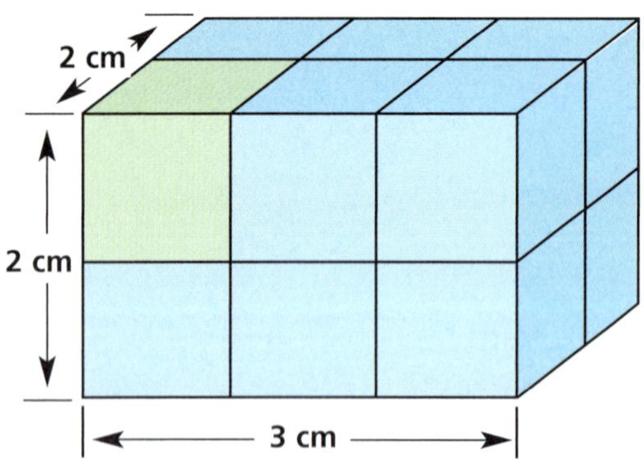

▲ **Figure 12** The volume of a box equals length times width times height.

$$V = l \times w \times h$$

If you have a box that is 10 cm on each side, its volume would be 1,000 cm^3. A liter is the same as 1,000 cm^3. One liter of liquid will fill the box exactly.

12 CALCULATE: How many milliliters of water would fill a 12-cm^3 box?

Hands-On Activity

CALCULATING AREA AND VOLUME

You will need 3 boxes of different sizes, paper, and a metric ruler.

1. Measure the length, width, and height of each box in centimeters. Record each measurement in your notes.
2. Calculate the volume of each box. Record each volume in your notes.
3. Find the surface area of each box. Record each area in your notes.

Practicing Your Skills

4. ANALYZE: Which of the three boxes has the largest volume?
5. CALCULATE: How many milliliters of liquid would fill each box?
6. ANALYZE: What is the surface area of the largest box?

Temperature **Temperature** is a measure of the amount of heat energy something contains. An instrument that measures temperature is called a thermometer.

Most thermometers are glass tubes. At the bottom of the tube is a wider part, called the bulb. The bulb is filled with liquid. Liquids that are often used include mercury, colored alcohol, or colored water. When heat is added, the liquid expands, or gets larger. It rises in the glass tube. When heat is taken away, the liquid contracts, or gets smaller. The liquid falls in the tube. On the side of the tube is a series of marks. You read the temperature by looking at the mark on the tube where the liquid stops.

Temperature can be measured on three different scales. These scales are the Fahrenheit (F) scale, the Celsius (C) scale, and the Kelvin (K) scale. The Fahrenheit scale is part of the English system of measurement. The Celsius scale is usually used in science. Almost all scientists, even in the United States, use the Celsius scale. Each unit on the Celsius scale is a degree Celsius (°C). The degree Celsius is the metric unit of temperature. Water freezes at 0°C. It boils at 100°C.

Scientists working with very low temperatures use the Kelvin scale. The Kelvin scale is part of the SI measurement system. It begins at absolute zero, or 0K. This number indicates, in theory at least, a total lack of heat.

COMPARING TEMPERATURE SCALES			
	Kelvin	Fahrenheit	Celsius
Boiling point of water	373K	212°F	100°C
Human body temperature	310K	98.6°F	37°C
Freezing point of water	273K	32°F	0°C
Absolute zero	0K	−459.67°F	−273.15°C

▲ **Figure 13**

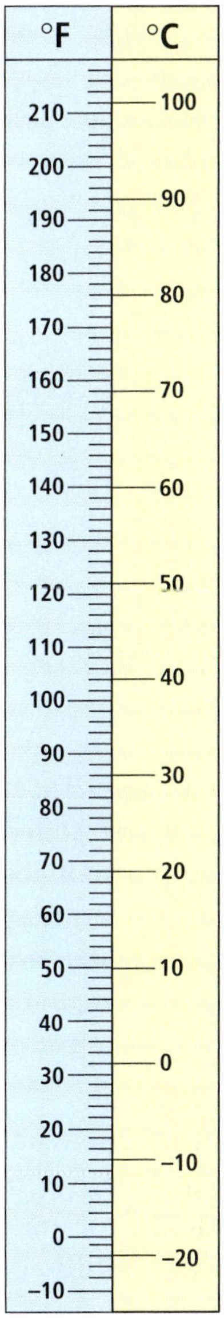

◀ **Figure 14** The Fahrenheit and Celsius scales

13 NAME: What are the three scales used to measure temperature?

Hands-On Activity

READING A THERMOMETER

You will need safety goggles, a lab apron, 2 beakers, a heat source, ice water, a wax pencil, a ruler, and a standard Celsius thermometer.

1. Boil some water in a beaker. ⚠CAUTION: Be very careful when working with heat. Place your thermometer in the beaker. Do not let the thermometer touch the sides or bottom of the beaker. Wait until the mercury rises as far as it will go. Record the temperature.
2. Fill a beaker with ice water. Place the unmarked thermometer into this beaker. Wait until the mercury goes as low as it will go. Record the temperature.

▲ **STEP 1** Record the temperature of the boiling water.

Practicing Your Skills

3. IDENTIFY: What is the temperature at which the mercury rose as high as it would go?
4. IDENTIFY: What is the temperature at which the mercury went as low as it would go?

5 Analyzing Data and Communicating Results

Key Term

communication: sharing information

Analyzing Data When you organize information, you put it in a logical order. In scientific experiments, it is important to organize your data. Data collected during an experiment are not very useful unless they are organized and easy to read. It is also important to organize your data if you plan to share the results of your experiment.

Scientists often organize information visually by using data tables, charts, graphs, and diagrams. By using tables, charts, graphs, and diagrams, scientists can display a lot of information in a small space. They also make it easier to compare and interpret data.

Tables are made up of rows and columns. Columns run up and down. Rows run from left to right. Tables display data in an orderly arrangement, often numerically. For example, reading a table containing the uses of sulfuric acid shows that the largest use of sulfuric acid is in fertilizers. Figure 15 is a table that shows some uses of sulfuric acid.

USES OF SULFURIC ACID	
Product	**Percentage**
Dyes, batteries, paint, explosives	15
Raw materials	15
Fertilizers	60
Petroleum refining	5
Metal processing	5

▲ **Figure 15**

Graphs, such as bar graphs, line graphs, and circle graphs, often use special coloring, shading, or patterns to represent information. Keys indicate what the special markings represent. Line graphs have horizontal (x) and vertical (y) axes to indicate such things as time and quantities.

14 EXPLAIN: How do tables and graphs help you analyze data?

Sharing Results When you talk to a friend, you are communicating, or sharing information. If you write a letter or a report, you are also communicating but in a different way. Scientists communicate all the time. They communicate to share results, information, and opinions. They write books and magazine or newspaper articles. They may also create Web sites about their work. This is called written **communication.**

Graphs are a visual way to communicate. The circle graph in Figure 16 is showing the same information that is shown in Figure 15. The circle graph presents the information in a different way.

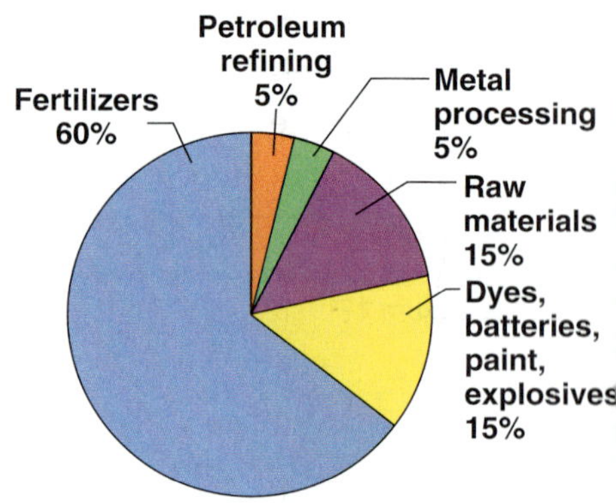

▲ **Figure 16** Circle graphs are a good way to show parts of a whole.

15 LIST: What are some ways to communicate the results of an experiment?

6 Making Predictions

Key Terms

infer: to form a conclusion

predict: to state ahead of time what you think is going to happen

Thinking of Possibilities When you **infer** something, you form a conclusion. This is called making an inference. Your conclusion will usually be based on observations or past experience. You may use logic to form your statement. Your statement might be supported by evidence and perhaps can be tested by an experiment. An inference is not a fact. It is only one possible explanation.

When you **predict,** you state ahead of time what you think will happen. Predictions about future events are based on inferences, evidence, or past experience. The two science process skills of inferring and predicting are very closely related.

16 CONTRAST: What is the difference between inferring and predicting?

How do you conduct a scientific investigation?

By now, you should have a good understanding of the science process skills. These skills are used to solve many science problems. There is also a basic procedure, or plan, that scientists usually follow when conducting investigations. Some people call this procedure the scientific method.

The scientific method is a series of steps that can serve as a guide to solving problems or answering questions. It uses many of the science process skills you know, such as observing and predicting.

Not all experiments use all of the steps in the scientific method. Some experiments follow all of them, but in a different order. In fact, there is no one right scientific method. Each problem is different. Some problems may require steps that another problem would not. However, most investigations will follow the same basic procedure.

1 DESCRIBE: What is the scientific method?

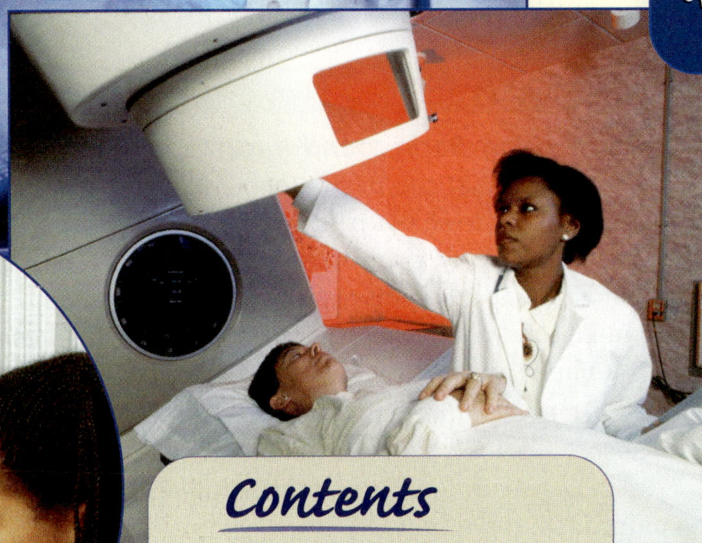

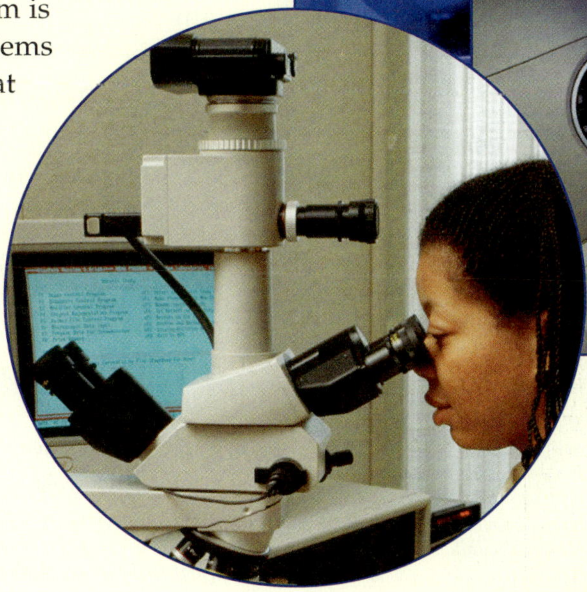

▲ **Figure 1** Scientists use the scientific method to guide experiments.

Contents

1 Identifying a Problem and Doing Research

Starting an Investigation Scientists often state a problem as a question. This is the first step in a scientific investigation. Most experiments begin by asking a scientific question. That is, they ask a question that can be answered by gathering evidence. This question is the reason for the scientific investigation. It also helps determine how the investigation will proceed.

Have you ever done background research for a science project? When you do this kind of research, you are looking for data that others have already obtained on the same subject. You can gather research by reading books, magazines, and newspapers, and by using the Internet to find out what other scientists have done. Doing research is the first step of gathering evidence for a scientific investigation.

2 **IDENTIFY:** What is the first step of a scientific investigation?

BUILDING SCIENCE SKILLS

Researching Background Information Suppose you notice that the brown paper towels at school do not seem to soak up as much water as the paper towels used in your home. You wonder which kinds of paper towels absorb, or soak up, the most water. You wonder if there is a connection between the paper towels' thickness and absorbency.

To determine which paper towels are most absorbent, look for information on paper towels in magazines, in books, or on the Internet. Put your findings in a report.

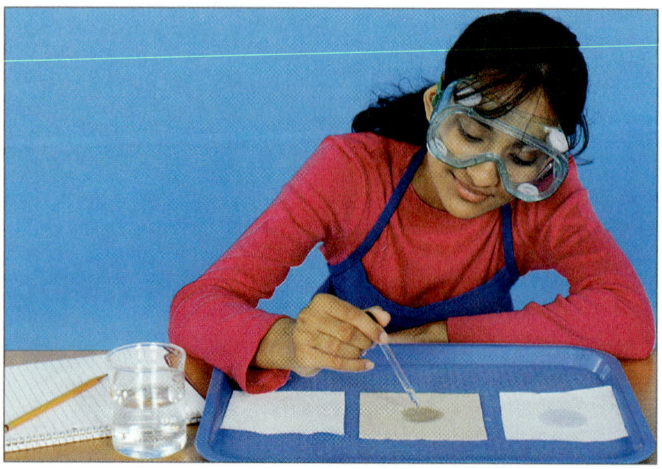

▲ **Figure 2** Testing paper towels for absorbency

2 Forming a Hypothesis

Key Terms

hypothesis: suggested answer to a question or problem

theory: set of hypotheses that have been supported by testing over and over again

Focusing the Investigation Scientists usually state clearly what they expect to find out in an investigation. This is called stating a hypothesis. A **hypothesis** is a suggested answer to a question or a solution to a problem. Stating a hypothesis helps to keep you focused on the problem and helps you decide what to test.

To form their hypotheses, scientists must think of possible explanations for a set of observations or they must suggest possible answers to a scientific question. One of those explanations becomes the hypothesis. In science, a hypothesis must include something that can be tested.

A hypothesis is more than just a guess. It must consider observations, past experiences, and previous knowledge. It is an inference turned into a statement that can be tested. A set of hypotheses that have been supported by testing over and over again by many scientists is called a **theory.** An example is the theory that explains how living things have evolved, or changed, over time.

A hypothesis can take the form of an "if . . . then" statement. A well-worded hypothesis is a guide for how to set up and perform an experiment.

3 **DESCRIBE:** How does a scientist form a hypothesis?

BUILDING SCIENCE SKILLS

Developing a Hypothesis A hypothesis for an experiment about which paper towels absorb the most water might be stated as follows:

If thicker paper towels soak up more water than thin paper towels, then thickness is an important factor for paper towel absorbency.

However, what do you mean by thicker? Are the paper towels really different? Does color or design make a difference? You need to make your hypothesis specific. Revise the hypothesis above to make it more specific.

3 Designing and Carrying Out an Experiment

Key Terms

variable: anything that can affect the outcome of an experiment

constant: something that does not change

controlled experiment: experiment in which all the conditions except one are kept constant

Testing the Hypothesis Scientists need to plan how to test their hypotheses. This means they must design an experiment. The plan must be a step-by-step procedure. It should include a record of any observations made or measurements taken.

All experiments must take variables into account. A **variable** is anything that can affect the outcome of an experiment. Room temperature, amount of sunlight, and water vapor in the air are just some of the many variables that could affect the outcome of an experiment.

4 **DEFINE:** What is a variable?

Controlling the Experiment One of the variables in an experiment should be what you are testing. This is what you will change during the experiment. All other variables need to remain the same. In this experiment, you will vary the type of paper towel.

A **constant** is something that does not change. If there are no constants in your experiment, you will not be sure why you got the results you did. An experiment in which all the conditions except one are kept constant is called a **controlled experiment.**

Some experiments have two setups. In one setup, called the control, nothing is changed. In the other setup, the variable being tested is changed. Later, the control group can be compared with the other group to provide useful data.

5 **EXPLAIN:** Explain how a controlled experiment is set up.

Designing the Procedure Suppose you want to design an experiment to determine if a paper towel's thickness affects its absorbency. You decide to do a set of measurements to find out the absorbency of three kinds of paper towels in a controlled environment. You will measure the thickness of the paper towels and then determine how much water each paper towel soaks up to see if your hypothesis is correct.

In designing your experiment, you need to identify the variables. The three kinds of paper towels are all variables that could affect the outcome of your experiment. Everything about testing the effect of thickness on absorbency needs to be the same except the actual thickness of each paper towel.

Finally, you should decide on the data you will collect. How will you measure the thickness of the paper towels? In this case, you might want to record the thickness of each towel, its color, whether it absorbed water, and how much water was absorbed.

The hands-on activity on page 12 is one possible experiment you could have designed. It has one method for measuring the absorbency of the paper towels. Sometimes scientists try to measure the same thing two different ways to be sure the test is accurate. Can you think of another method to measure the absorbency of paper towels?

6 **LIST:** How do constants and variables affect an experiment?

Hands-On Activity

CARRYING OUT AN EXPERIMENT

You will need 3 or more kinds of paper towels, a metric ruler, an eyedropper or pipette (preferably calibrated in millimeters), and water.

1. Get three different kinds of paper towels. To find the thickness of each kind of towel, measure the thickness of five towels and divide the result by five. Set up a data table for the information you gather.
2. Now you are ready to compare the absorbency of the paper towels. Cut the paper towels into squares of equal sizes. Squares that are 10 cm in size are good for testing.
3. Lay a square of paper towel on a tray or other nonabsorbent surface. Add drops of water one at a time until the paper towel has soaked up all the water it can. Record how much water was absorbed. If your eyedropper is not marked in milliliters, you can record your data in "drops." If it is marked, then you should record the milliliters.
4. Test all the paper towel samples the exact same way. Be sure that you measure only the water that is absorbed and that you let each paper towel sample soak up as much water as it will hold.
5. You are now ready to compare your data and see if they support your hypothesis.

Practicing Your Skills

6. **OBSERVE:** What happened in the experiment? How much water did each paper towel sample absorb?
7. **COMPARE:** Which paper towel absorbed the most?
8. **EXPLAIN:** What procedures did you follow to make sure the paper towels were all given a fair and equal test?
9. **IDENTIFY:** What is the variable being tested in this experiment?

4 Recording and Analyzing Data

Dealing With Data During an experiment, you must keep careful notes about what you observe. For example, you might need to note any special steps you took in setting up the experiment, exactly how you made the drops the same size each time, or the temperature of the water. This is important information that might affect your conclusion.

At the end of an experiment, you will need to study the data to find any patterns. Much of the data you will deal with is written text. You may read a report or a summary of an experiment. However, scientific information is often a set of numbers or facts presented in other, more visual ways. These visual presentations make the information more meaningful and easier to understand. Tables, charts, and graphs, for instance, help you understand a collection of facts on a topic.

After your data have been organized, you need to ask what the data show. Do they support your hypothesis? Do they show something wrong in your experiment? Do you need to gather more data by performing another experiment?

7 **LIST:** What are some ways to display data?

BUILDING SCIENCE SKILLS

Analyzing Data You made the following notes during your experiment. How would you display this information?

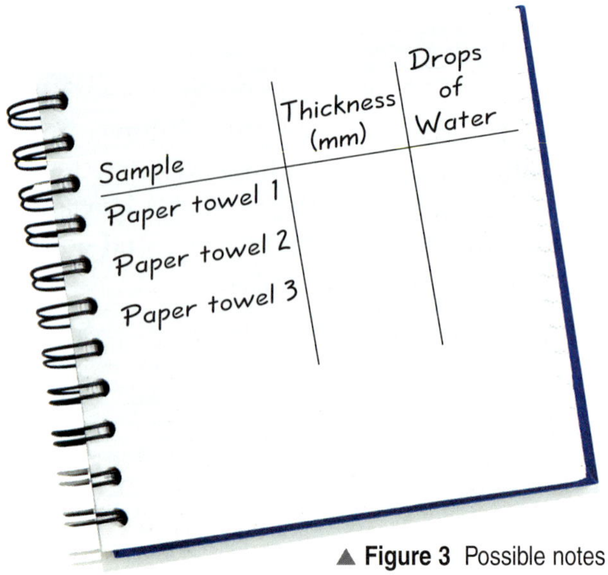

▲ **Figure 3** Possible notes

5 Stating a Conclusion

Drawing Conclusions A conclusion is a statement that sums up what you have learned from an experiment. When you draw a conclusion, you need to decide whether the data you collected supported your hypothesis. You may need to repeat an experiment several times before you can draw any conclusions from it. Conclusions often lead you to ask new questions and plan new experiments to answer them.

8 **EXPLAIN:** Why might it be necessary to repeat an experiment?

BUILDING SCIENCE SKILLS

Stating a Conclusion Review your hypothesis statement regarding the effect of the thickness of a paper towel on its absorbency. Then, review the data you obtained during your experiment.

- Was your hypothesis correct? Use your observations to support your answer.
- Which paper towel absorbed the most? Was it also the thickest?

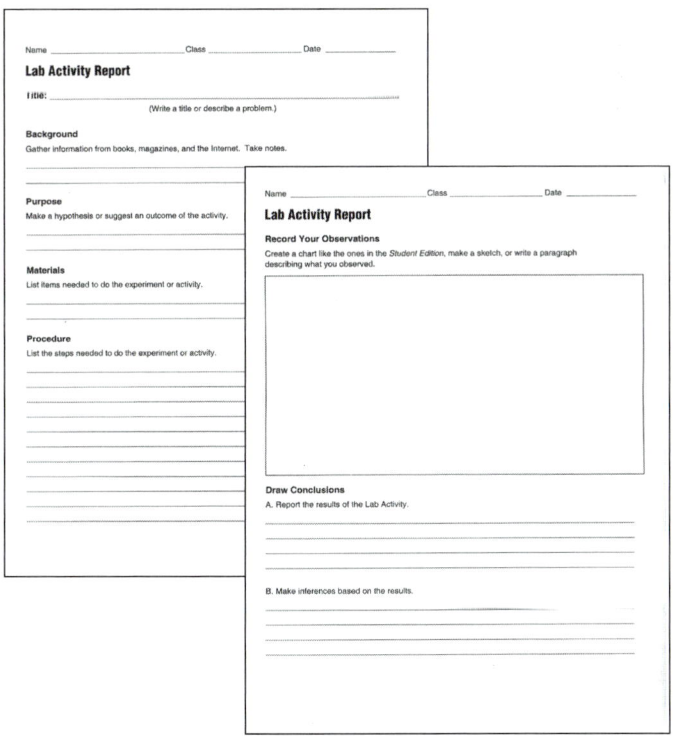

Name ____ Class ____ Date ____

Lab Activity Report

Title: ____

(Write a title or describe a problem.)

Background

Gather information from books, magazines, and the Internet. Take notes.

Purpose

Make a hypothesis or suggest an outcome of the activity.

Materials

List items needed to do the experiment or activity.

Procedure

List the steps needed to do the experiment or activity.

Name ____ Class ____ Date ____

Lab Activity Report

Record Your Observations

Create a chart like the ones in the *Student Edition*, make a sketch, or write a paragraph describing what you observed.

Draw Conclusions

A. Report the results of the Lab Activity.

B. Make inferences based on the results.

▲ **Figure 4** Throughout this program, you may use forms like these to organize your lab reports.

6 Writing a Report

Communicating Results Scientists keep careful written records of their observations and findings. These records are used to create a lab report. Lab reports are a form of written communication. They explain what happened in the experiment. A good lab report should be written so that anyone reading it can duplicate the experiment. It should contain the following information:

- A title
- A purpose
- Background information
- Your hypothesis
- Materials used
- Your step-by-step procedure
- Your observations
- Your recorded data
- Your analysis of the data
- Your conclusions

Your conclusions should relate back to the questions you asked in the "purpose" section of your report. Also, the report should not have any experimental errors that might have caused unexpected results. For example, did you follow the steps in the correct order? Did an unexpected variable interfere with your results? Was your equipment clean and in good working order? This explanation of possible errors should also be part of your conclusions.

9 **EXPLAIN:** Why is it important to explain possible errors in your lab report?

BUILDING SCIENCE SKILLS

Writing a Lab Report Write a lab report to communicate to other scientists your discoveries about the effect of a paper towel's thickness on its water absorbency. Your lab report should include a title, your hypothesis statement, a list of materials you used, the procedure, your observations, and your conclusions. Try to include one table of data in your report.

LAB SAFETY

Working in a science laboratory can be both exciting and meaningful. However, you must always be aware of safety precautions when carrying out experiments. There are a few basic rules that should be followed in any science laboratory:

- Read all instructions carefully before the start of an experiment. Follow all instructions exactly and in the correct order.
- Check your equipment to make sure it is clean and working properly.
- Never taste, smell, or touch any substance in the lab that you are not told to do so. Never eat or drink anything in the lab. Do not chew gum.
- Never work alone. Tell a teacher at once if an accident occurs.

Experiments that use chemicals or heat can be dangerous. The following list of rules and symbols will help you avoid accidents. There are also rules about what to do if an accident does occur. Here are some rules to remember when working in a lab:

1. Do not use glass that is chipped or metal objects with broken edges. Do not try to clean up broken glassware yourself. Notify your teacher if a piece of glassware is broken.
2. Do not use electrical cords with loose plugs or frayed ends. Do not let electrical cords cross in front of working areas. Do not use electrical equipment near water.
3. Be very careful when using sharp objects such as scissors, knives, or tweezers. Always cut in a direction away from your body.
4. Be careful when you are using a heat source. Use proper equipment, such as tongs or a ringstand, when handling hot objects.
5. Confine loose clothing and hair when working with an open flame. Be sure you know the location of the nearest fire extinguisher. Never reach across an open flame.
6. Be careful when working with poisonous or toxic substances. Never mix chemicals without directions from your teacher. Remove any long jewelry that might hang down and end up in chemicals. Avoid touching your eyes or mouth when working with these chemicals.
7. Use extreme care when working with acids and bases. Never mix acids and bases without direction from your teacher. Never smell anything directly. Use caution when handling chemicals that produce fumes.
8. Wear safety goggles, especially when working with an open flame, chemicals, and any liquids.
9. Wear lab aprons when working with substances of any sort, especially chemicals.
10. Use caution when handling or collecting plants. Some plants can be harmful if they are touched or eaten.
11. Use caution when handling live animals. Some animals can injure you or spread disease. Handle all live animals as humanely as possible.
12. Dispose of all equipment and materials properly. Keep your work area clean at all times.
13. Always wash your hands thoroughly with soap and water after handling chemicals or live organisms.
14. Follow the ⚠ CAUTION and safety symbols you see used throughout this book when doing labs or other activities.

Chapter 1 Heat

▲ **Figure 1-1** Because hot air rises, this balloon can fly.

Imagine floating high above Earth's surface in a hot-air balloon like the one in Figure 1-1. In order to keep the balloon high above the ground, air in the balloon must be heated. Heat energy from the flames causes the particles that make up air to gain energy. The particles move faster and spread farther apart. As a result, the balloon expands and rises high above Earth's surface.

▶ What do you think would happen if the air in the balloon was cooled?

Contents

1-1 What is heat?

INVESTIGATE

Observing Heat

HANDS-ON ACTIVITY

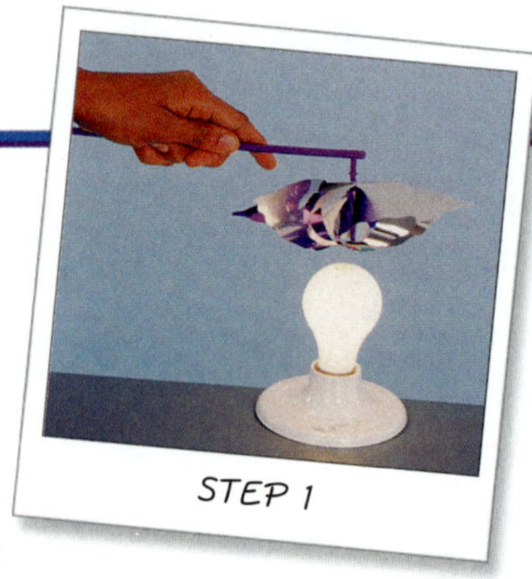

1. Hold a pinwheel over a light source that is not lit. Observe the action of the pinwheel for a full minute.
2. Turn on the light source and hold the pinwheel over it. Observe the action of the pinwheel for a full minute.

THINK ABOUT IT: Compare the action of the pinwheel in Steps 1 and 2. Did the light have any effect on the pinwheel? What type of energy is coming from the light?

Objective

Explain how heat is a form of energy.

Key Term

heat: energy of particles moving from warmer regions to cooler regions

Caloric Have you ever walked on a sandy beach on a hot, summer day? If you have, you probably know that the burning feeling in your feet was caused by the heat in the sand. At one time, people thought that heat was a physical substance. They called this substance caloric (kuh-LAWR-ihk). Caloric was thought to flow like a liquid. Today, we know that heat is not a substance. It is a form of energy. The heat in the sand actually comes from energy absorbed from the Sun.

1 **EXPLAIN:** What is caloric?

Moving Particles You have learned that matter is made up of tiny particles called atoms and molecules. These tiny particles are always in motion. For example, a brick within a wall is motionless, but the tiny particles that make up the brick are constantly moving. The motion of these particles results in heat, or thermal energy.

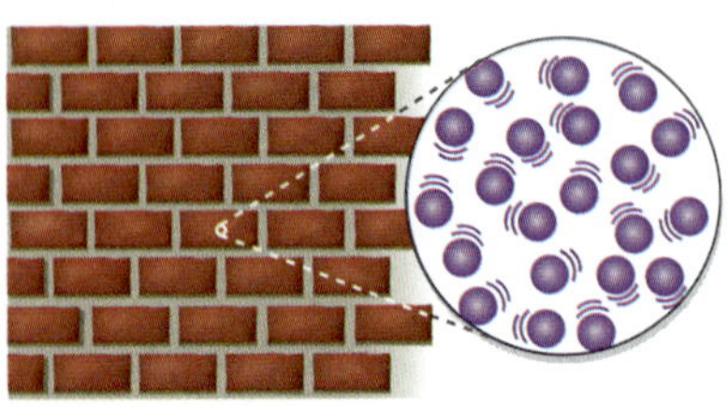

◀ **Figure 1-2** The tiny particles that make up a brick are constantly moving.

Heat is the energy of particles moving from a warmer region to a cooler region. All matter has heat. Even an ice cube contains heat, because the molecules of water that make up the ice are moving back and forth. If heat energy is added to the ice cube, the molecules move faster and farther apart. If enough heat energy is added to the ice, the molecules will gain enough energy to leave the solid state and become a liquid. With continued heating, the molecules will gain enough energy to break away from the liquid state and become a gas.

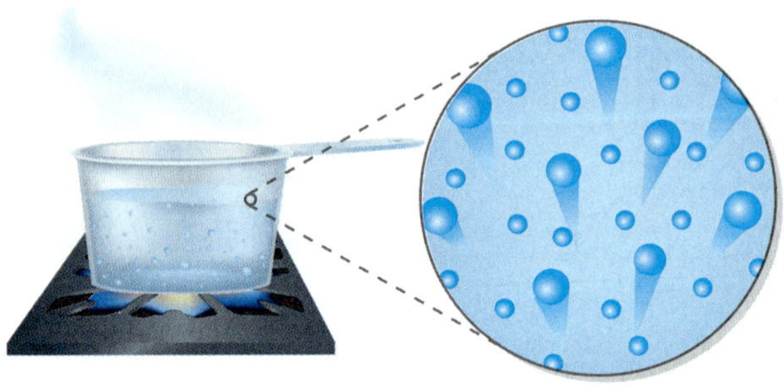

▲ **Figure 1-3** Adding heat energy to water makes the water molecules move faster and farther apart.

2 **DEFINE:** What is heat?

Movement of Heat What happens when you hold a glass of ice water in your hand? You may notice that after a few minutes, the ice in the water begins to melt. At the same time, your hand gets cold. This happens because heat energy moves from a warmer object to a cooler object. Your hand is warmer than the ice water. Heat from your hand moved into the glass. As heat entered the glass, the water and the ice absorbed the heat and became warmer. Your hand lost some heat energy so it began to feel cooler.

3 **DESCRIBE:** In what direction does heat move?

Heat and Work How do scientists know that heat is a form of energy? Energy can do work. Remember that something must be moved for work to be done. When you boil water, you can see the water bubbling and moving inside the container. Heat is doing work.

Heat does work when it causes a rocket to be lifted into the air. Heat causes the fuel in a rocket engine to explode, producing hot gases. These hot gases expand rapidly. The expanding gases rush out of the rocket engine. As they do, they push against the rocket, forcing it upward. The moving rocket shows that heat is a form of energy that can do work.

▲ **Figure 1-4** Expanding hot gases cause the rocket to move upward.

4 **INFER:** How do scientists know that heat is a form of energy?

CHECKING CONCEPTS

1. At one time, people thought that heat was a substance called ___________.
2. Moving particles have ___________ energy.
3. The energy of particles moving from warmer regions to cooler regions is ___________.
4. Adding heat energy makes particles of matter move ___________.
5. For work to be done, something must be ___________.

THINKING CRITICALLY

6. **INFER:** Explain how a piece of wood has kinetic energy.
7. **ANALYZE:** Suppose you hit a piece of metal several times with a hammer. When you touch the piece of metal, it feels hot. Explain why the metal gets hot after being hit with the hammer.
8. **HYPOTHESIZE:** Explain what will happen when you place an ice cube in a cup of hot coffee.

Web InfoSearch

Seeing Heat Because heat is energy, you usually cannot see it. Sometimes, however, you can see evidence of it. For example, when steam rises over a pot of boiling water, you can "see" the motion of the heat. Also, wavy air over a hot road is caused by heat. Scientists have invented a technology to see heat. Thermograms show the heat coming from a person or an object. This information is used in many ways.

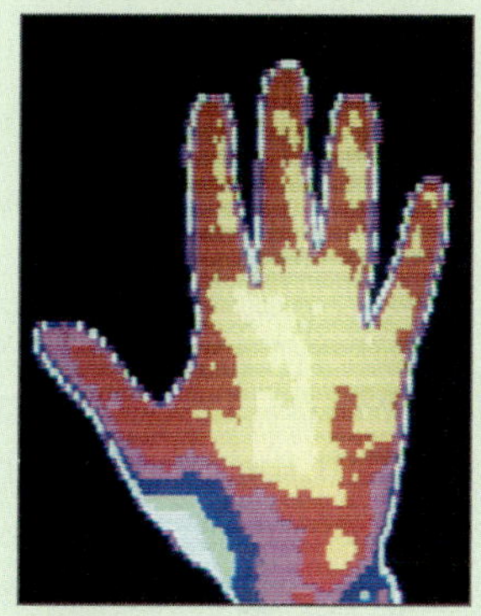

▲ **Figure 1-5** This thermogram shows the heat coming off a human hand.

SEARCH: Use the Internet to find out more about thermograms. What information might a thermogram of a house or office building tell you? Start your search at www.conceptsandchallenges.com. Some key search words are **thermogram heat loss, energy audit thermogram,** and **thermogram building envelope.**

1-2 What is temperature?

INVESTIGATE

Observing Temperature Differences
HANDS-ON ACTIVITY

1. Fill a pan with cold water. Fill a second pan with warm water. Fill a third pan with hot but not boiling water.
2. Put both of your hands in the warm water. Observe how your hands feel. Dry your hands. Put one hand into the cold water and the other hand into the hot water.
3. After 1 minute, put both of your hands into the warm water.

THINK ABOUT IT: Did the temperature of the warm water feel the same in both of your hands?

Objective
Differentiate between heat and temperature.

Key Terms
temperature: measure of the average kinetic energy of the particles in a sample of matter

calorie: unit of heat; amount of heat needed to raise the temperature of 1 g of water 1°C

Temperature The measure of the average kinetic energy of all the particles in a sample of matter is called **temperature.** If you compared a bowl of hot soup to a bowl of ice cream, you would find that the average kinetic energy of the soup's particles is greater than the average kinetic energy of the ice cream's particles. The temperature of the soup is higher than the temperature of the ice cream.

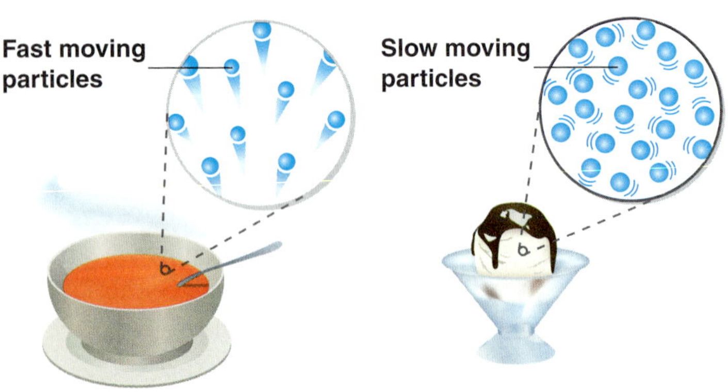

▲ **Figure 1-6** The average kinetic energy of the particles in the soup is greater than the average kinetic energy of the particles in the ice cream.

1 DEFINE: What is temperature?

Heat and Temperature Heat is related to temperature, but they are not the same. Heat is the *total* kinetic energy of all the particles in a sample of matter. Temperature is the measure of the *average* kinetic energy of all the particles in a sample of matter.

Two samples of the same material may have the same temperature but not the same amount of heat. For example, a pail of ocean water may have a temperature of 27°C. The ocean water near the beach may also have a temperature of 27°C. However, because there are fewer particles of ocean water in the pail, the amount of heat in it is less than the amount of heat in the nearby ocean.

▲ **Figure 1-7** The pail of ocean water has fewer particles in it than the ocean does. So, at the same temperature, the pail of ocean water has less heat than the ocean.

2 COMPARE: At the same temperature, which has more heat, the water in a bucket or the water in an ocean?

Measuring Heat Heat is directly related to the temperature of a sample of matter. When heat is added to a sample of matter, the average kinetic energy of the particles of matter increases. So, the temperature of the sample goes up. The opposite is also true. When heat is removed from a sample of matter, the average kinetic energy of the particles of matter decreases. The temperature of the sample goes down.

The terms "hot" and "cold" are often used to describe how much heat energy a sample of matter contains. It is important to understand that these terms are not measures of heat energy. They are used to compare the temperature of one sample of matter to that of another.

Heat is measured by describing the temperature change it causes. Heat is measured in units called calories. A **calorie** is the amount of heat needed to raise the temperature of 1 gram of water 1°C. One gram of water is a small amount of water. So, a calorie is a small amount of heat.

3 PREDICT: What happens when you add heat energy to a substance?

CHECKING CONCEPTS

1. Two samples of the same substance may have the same temperature but not contain the same amount of __________.
2. When heat is removed from water, its temperature __________.
3. Heat affects the __________ of a substance.
4. A __________ is a unit of heat.

THINKING CRITICALLY

5. COMPARE: Compare the average kinetic energy of the particles in a cup of hot tea to the particles in a glass of iced tea.

DESIGNING AN EXPERIMENT

Design an experiment to solve the following problem. Include a hypothesis, variables, a procedure, and a type of data to study.

PROBLEM: Many people think that coldness can be added to a substance. How can you demonstrate that coldness is an absence of heat?

How Do They Know That?

HEAT IS A FORM OF ENERGY

Benjamin Thompson was born in Woburn, Massachusetts. During the Revolutionary War, Thompson remained loyal to Great Britain and acted as a spy for the British. During the war, Thompson fled the colonies to live in Great Britain. He was knighted by King George III in 1784. Seven years later, Thompson was made a count of the Holy Roman Empire.

Thompson was also a scientist. Around 1798, Thompson was in charge of a factory that made cannons. The machines used to drill the cannons were turned by horses. Thompson observed that the cannons became very hot as they were drilled. He decided to try an experiment. He surrounded a cannon with a box filled with water. As the cannon was drilled, the water began to boil. The water continued boiling as long as the drilling went on. At that time, people believed that heat was a substance called caloric. Thompson concluded that the heat was produced by the motion of the drill, not a flow of caloric. This was the first step in showing that heat is a form of energy and not a substance.

▲ **Figure 1-8** Benjamin Thompson showed that heat is not a flow of caloric.

Thinking Critically What do you think happened to the water once the drilling stopped?

1-3 What is freezing point?

Objective

Identify the freezing point and melting point of a substance.

Key Terms

freezing point: temperature at which a liquid changes to a solid

melting point: temperature at which a solid changes to a liquid

Freezing Water When water freezes, it changes to ice. Suppose you put a beaker of water into a freezer whose temperature is set at –10°C. As the water loses heat, its temperature will drop. When the temperature reaches 0°C, the water will begin to freeze. At this point, the temperature of the water stops going down. As the water changes to ice, more heat is lost. After a while, the water will completely change to ice.

The temperature at which a liquid turns to ice is called its **freezing point.** The freezing point of water is 0°C. Once the water has completely changed to ice, the temperature of the ice can drop below 0°C. Figure 1-9 shows how the temperature of water changes as it freezes.

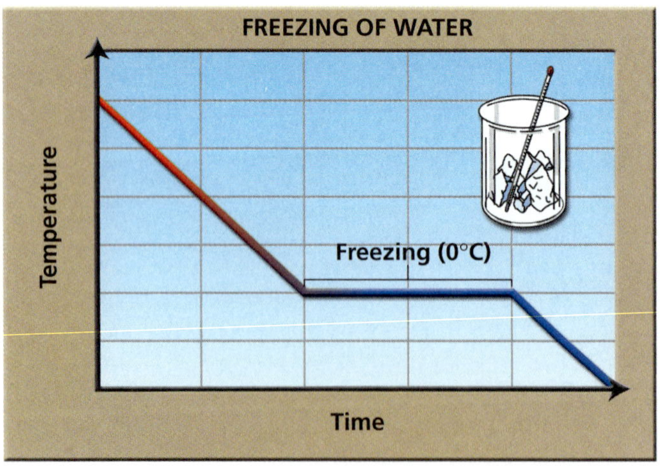

▲ **Figure 1-9** When water reaches its freezing point, it stays at 0°C until it completely changes to ice. Its temperature can then fall below 0°C. The temperature will stop falling once it equals its surroundings.

1 DEFINE: What is the freezing point of water?

Freezing Points of Other Liquids When heat is removed from a liquid, its temperature goes down. Removing heat from a liquid at its freezing point changes the liquid into a solid. Every liquid has its own freezing point. Figure 1-10 shows the freezing points of some liquids.

FREEZING POINTS OF SOME LIQUIDS	
Liquid	**Freezing point**
Acetic acid	17°C
Acetone	–95°C
Benzene	6°C
Ethyl alcohol	–117°C
Glycerine	18°C
Mercury	–39°C
Water	0°C

▲ **Figure 1-10**

2 STATE: What is the freezing point of ethyl alcohol?

Absolute Zero As you remove heat from a sample of matter, the particles of the matter slow down. The temperature of the sample falls. If you continue to remove heat from the sample, you might expect to reach a point at which the particles of matter stop moving. At such a point, no more heat could be removed from the sample of matter. The temperature of the sample would be absolute zero.

Absolute zero is described as the coldest temperature possible. Its value is about –273°C! In order to lower the temperature of something, its heat must be transferred to something colder. Because absolute zero is the coldest possible temperature, it will never be reached.

3 DESCRIBE: What happens to the particles of a substance as the temperature nears absolute zero?

Melting Point The temperature at which a solid changes to a liquid is called its **melting point.** The freezing point and the melting point of a substance can be the same. When heat is removed from a liquid, it changes to a solid. When heat is added to a solid, it changes back to a liquid. The melting point of ice is 0°C—the same as its freezing point.

▲ **Figure 1-11** The freezing point and the melting point of water are the same, 0°C.

4 **STATE:** What is the melting point of water?

CHECKING CONCEPTS

1. What happens when water freezes?
2. At what temperature does water begin to freeze?
3. What is the temperature called at which all particle motion almost stops?
4. What is the melting point of ice?

THINKING CRITICALLY

5. **HYPOTHESIZE:** The water in a lake has just reached 0°C. Explain why it may not be safe to go ice skating on the lake.
6. **COMPARE:** How are melting point and freezing point different? How are they alike?

INTERPRETING VISUALS

Look back at Figure 1-9 to answer the following questions.

7. **INFER:** What does the red part of the graph indicate?
8. **ANALYZE:** What happens to the temperature of water as time passes?
9. **STATE:** What happens to the temperature of water after it reaches the freezing point?

Science and Technology

CRYOGENICS

When heat energy is removed from a substance, the temperature of the substance goes down. When all of the available heat has been removed from a substance, its temperature cannot go down any further. At this temperature, all the particles of the substance almost stop moving. This is the lowest possible temperature that can be reached. It is called absolute zero. This temperature has never been reached. However, scientists have been able to cool substances to within 0.01°C of absolute zero. The study of such very low temperatures is called cryogenics (kriy-oh-JEHN-ihks).

Cryogenics is important in the refrigeration of food, in space technology, and in medicine. In cryogenic surgery, a surgeon uses an extremely cold probe instead of a scalpel. Cryogenic surgery reduces bleeding. It can also reduce much of the pain felt after the surgery. Many cryogenic operations can be done in a doctor's office.

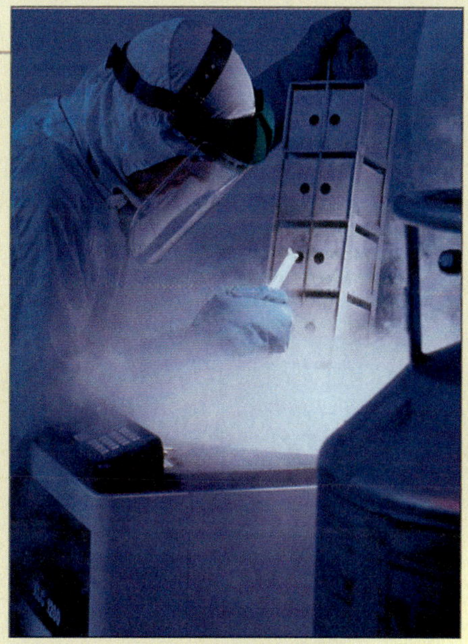

▲ **Figure 1-12** In a cryogenics laboratory, substances are cooled to very low temperatures.

Cryogenics may help the automobile industry. Hydrogen and oxygen are gases at normal temperatures. However, at very low temperatures, they become liquids. Liquid hydrogen and liquid oxygen are used as fuels in some rockets. In the future, liquid hydrogen may replace gasoline as a fuel for automobiles.

Thinking Critically How has cryogenics helped different industries?

1-4 What is boiling point?

Objectives

Identify the boiling point of a liquid. Differentiate between boiling and evaporation.

Key Terms

boiling point: temperature at which a liquid changes to a gas

evaporation (ee-vap-uh-RAY-shuhn)**:** change from a liquid to a gas at the surface of the liquid

Boiling Water When water boils, it changes to steam. Steam is water in the form of a gas. As you heat water, the temperature of the water rises. As the temperature of the water rises, small bubbles appear in the water. These bubbles show that a gas, steam, is being formed. As you continue to heat the water, more and more bubbles are formed. At 100°C, the bubbles start to leave the container. However, the temperature of the water stays at 100°C. The temperature remains at 100°C until all of the water has changed to steam.

The temperature at which a liquid changes to a gas is called its **boiling point.** The boiling point of water is 100°C. After all the water turns into steam, its temperature can then rise above 100°C. Figure 1-13 shows how the temperature of water changes as it boils.

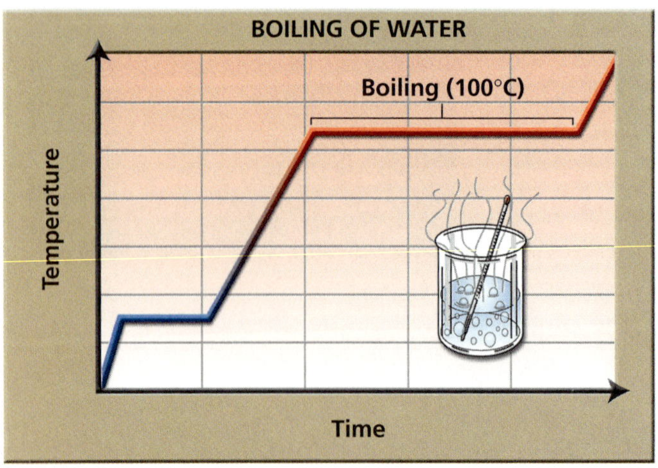

▲ **Figure 1-13** When water reaches its boiling point, it stays at 100°C until it completely changes to steam. The temperature of the steam can then rise above 100°C.

1 DEFINE: What is boiling point?

Boiling Points of Other Liquids When heat is added to a liquid, the temperature of the liquid rises. When the liquid reaches its boiling point, it begins to form a gas. Every liquid has its own boiling point. Figure 1-14 shows the boiling points of some liquids.

BOILING POINTS OF SOME LIQUIDS	
Liquid	**Boiling point**
Acetic acid	118°C
Acetone	57°C
Benzene	80°C
Ethyl alcohol	78°C
Glycerine	290°C
Mercury	357°C
Water	100°C

▲ **Figure 1-14**

2 ANALYZE: Which liquid listed in Figure 1-14 has the highest boiling point?

Evaporation When a liquid is allowed to stand uncovered at room temperature, it slowly changes to a gas. This change is called **evaporation.** Evaporation happens only at the surface of a liquid and can occur at temperatures below the boiling point. When a liquid evaporates, some particles at the surface of the liquid gain enough energy to escape into the air. This is different from when a liquid boils. When a liquid boils, bubbles form throughout the liquid and rise out of the liquid.

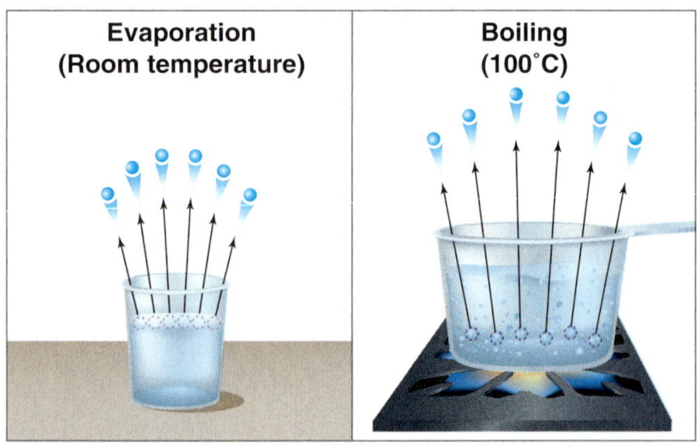

▲ **Figure 1-15** Evaporation occurs at the surface of a liquid (left). When a liquid boils, gas bubbles form throughout the liquid and then rise out of the liquid (right).

3 DEFINE: What is evaporation?

CHECKING CONCEPTS

1. When water __________, it changes to steam.
2. The temperature at which water changes to steam is its __________.
3. The boiling point of water is __________°C.
4. When a liquid reaches its boiling point, it begins to change to a __________.
5. An uncovered liquid changes to a gas at room temperature by the process of __________.
6. Evaporation occurs only at the __________ of a liquid.

THINKING CRITICALLY

7. **COMPARE:** In your own words, explain the difference between boiling and evaporation.
8. **HYPOTHESIZE:** Why is it possible to get a more serious burn from steam than from boiling water?
9. **EXPLAIN:** As water is heated, what evidence is there that a gas is forming?
10. **INFER:** When does the temperature of boiling water rise above 100°C?

Web InfoSearch

Boiling Point and Elevation The boiling point of liquids can change depending upon elevation above or below sea level. At elevations above sea level, the boiling point is lower. At elevations below sea level, the boiling point is higher.

SEARCH: Use the Internet to find the boiling point of water at different elevations above and below sea level. Why does elevation affect the boiling point of water? Start your search at www.conceptsandchallenges.com. Some key search words are **water boiling point altitude** and **water boiling point elevation.**

Hands-On Activity

CHANGING THE BOILING POINT OF WATER

You will need three beakers, water, a measuring cup, a balance, salt, a stirrer, a thermometer, and a hot plate.

1. Pour 100 mL of water into each beaker.
2. Add 10 g of salt to the first beaker. Add 20 g of salt to the second beaker. Do not add any salt to the third beaker. Stir the salt in the first and second beakers.
3. Heat the water in each beaker until it boils. Record the temperature of the water in each beaker as it boils.

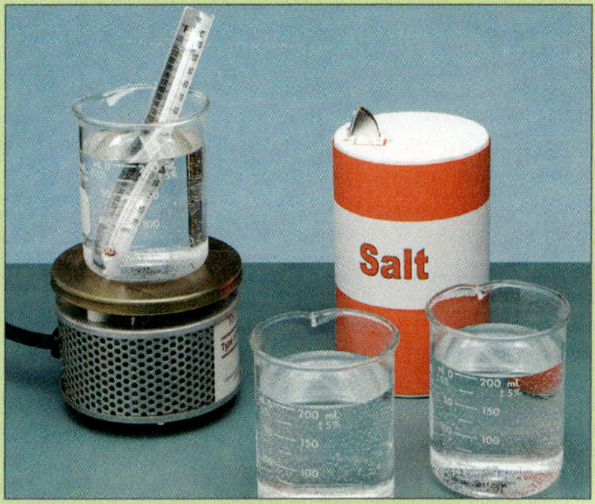

▲ **STEP 3** Record the temperature of the water as it boils.

Practicing Your Skills

4. **OBSERVE:** What is the boiling point of the water in the first beaker? The second beaker? The third beaker?
5. **INFER:** How does adding 10 g of salt affect the boiling point of water?
6. **INFER:** How does adding 20 g of salt affect the boiling point of water?

1-5 What is conduction?

INVESTIGATE

Observing Heat Transfer

HANDS-ON ACTIVITY

1. Half-fill a jar with hot water. Place a wooden spoon, a metal spoon, and a plastic spoon into the jar. Do not allow the spoons to touch each other.
2. After a full minute, touch the handle of each spoon.

THINK ABOUT IT: Did the handles of all three spoons feel the same? If not, how can you explain the difference?

STEP 1

Objective

Describe how heat is transferred through solids.

Key Terms

conduction (kuhn-DUK-shuhn)**:** process of heat transfer in solids

conductor: material that conducts heat easily

insulator: material that does not conduct heat easily

Heat Transfer in Solids A blacksmith hammers a piece of iron to make horseshoes. When a blacksmith places a piece of iron into a flame, the iron gets very hot. The heat energy causes the particles in the iron to bump into each other. After a while, the iron gets hot enough so that it becomes soft. The piece of iron can then be hammered into shape.

▲ **Figure 1-16** As the iron gets hot, the particles within it bump into each other.

Heat is transferred through solids by **conduction.** You learned that heat moves from an object with a higher temperature to an object with a lower temperature. When fast-moving particles in a sample of matter bump into slow-moving particles, heat energy passes from the fast-moving particles to the slow-moving particles. As a piece of iron is heated, heat energy is transferred from the fast-moving particles to the slow-moving particles. The slow-moving particles gain energy and bump into other particles. In this way, heat energy is transferred throughout the sample of matter.

1 **DEFINE:** What is conduction?

Conductors of Heat All metals are good conductors of heat. A **conductor** is a material that allows heat to move through it easily. Copper, gold, silver, iron, aluminum, and steel are all good conductors of heat. Copper and silver are two of the best conductors of heat. Heat will travel faster through items made of copper and silver than it will through items made of iron or steel.

2 **IDENTIFY:** What substances are good conductors of heat?

Poor Conductors of Heat Many materials are poor conductors of heat. Materials that do not conduct heat easily are called **insulators.** Wood, paper, wax, and air are poor conductors of heat. These materials are insulators.

Insulators prevent heat from moving from place to place. Houses are insulated to keep them warm in winter and cool in summer. To insulate houses, spaces are left between the inside and outside walls of the house. The spaces are filled with an insulating material. During winter, this insulation helps keep heat from escaping to the outside. During the summer, insulation helps keep heat from getting into the house.

▲ **Figure 1-17** Pots made of copper are good conductors of heat. Handles made of plastic are good insulators.

3 **EXPLAIN:** Why are insulators used in houses?

CHECKING CONCEPTS

1. How does heat travel by conduction?
2. Name two materials that are good conductors of heat.
3. What is an insulator?
4. Name two materials that are good insulators.

THINKING CRITICALLY

5. **INFER:** Explain how food in a pot gets heated by conduction.
6. **EXPLAIN:** How is a blanket an insulator?

BUILDING SCIENCE SKILLS

Researching Insulators do not allow heat to move easily through them. Look throughout your home to find examples of insulators. Make a list of them and describe what kind of material they are made of. Explain how each one of them acts as an insulator.

Real-Life Science

HOME INSULATION

▲ **Figure 1-18** Home insulation keeps houses warm in winter and cool in summer.

In many parts of the United States, winters are cold and summers are hot. In these parts of the country, houses must be insulated. Home insulation keeps houses warm in winter and cool in summer. Home insulation can help reduce the amount of fuel needed to heat or cool a home. Good insulation can cut fuel use by as much as 50%.

Insulation is needed in those parts of a house where the most heat is usually lost. In most homes, heat loss occurs through the attic floor, the ceiling of an unheated basement, and the side walls. Different types of insulation can be used in these places. For example, blankets of fiberglass can be inserted between beams in floors and ceilings. Liquid plastic foam can be sprayed into the spaces between the inside and outside walls.

A number called an R-value is used to grade insulating materials. An insulating material with a high R-value is best at preventing heat loss. Choosing insulation with the highest R-value can greatly reduce fuel costs.

Thinking Critically How does insulation keep a house cool in summer?

1-6 What is convection?

Objective

Describe how heat travels through gases and liquids.

Key Terms

convection (kuhn-VEHK-shuhn)**:** transfer of heat in gases and liquids

convection current: movement of gases or liquids caused by differences in density

Heat Transfer in Gases and Liquids The transfer of heat in gases and liquids is called **convection.** When gases and liquids are heated, the particles gain energy and move faster and farther apart. Because these particles end up farther apart, the heated gas or liquid becomes less dense than the cooler, surrounding gas or liquid. This makes the less-dense gas or liquid rise and the cooler, denser gas or liquid moves in to take its place. As it rises, the gas or liquid carries heat with it.

1 DEFINE: What is convection?

Convection Currents Heat is carried in gases and liquids by means of convection currents. **Convection currents** are movements of gases or liquids caused by differences in density. Convection currents transfer heat. Air is a gas. Air at Earth's surface is warmed by the Sun and then rises. Cool air above the North Pole and the South Pole sinks. The rising of warm air and the sinking of cool air cause convection currents. Certain birds, such as eagles and hawks, use rising warm air in convection currents to soar above Earth's surface.

▲ **Figure 1-19** An eagle uses convection currents to soar.

Convection currents are found in liquids as well as in air. In bodies of water, warmer water rises to the surface and carries heat with it. Cooler water sinks to the bottom. Convection currents cause heat to be transferred through bodies of water. When water is heated in a pot, convection currents form. Figure 1-20 shows convection currents in a pot of boiling water.

◀ **Figure 1-20** Convection currents transfer heat in a liquid.

Just below Earth's crust is a layer of rock called the mantle. The mantle is heated from below by heat from Earth's core. Rock in the upper part of the mantle behaves like a very thick liquid. Scientists believe that convection currents traveling through the upper mantle are responsible for the very slow movement of large sections of Earth's crust. This concept is known as the theory of plate tectonics.

2 EXPLAIN: How do convection currents transfer heat in liquids?

Uses of Convection Heat transfer by convection is used in some home heating systems. In a hot-water heating system, water is heated in a hot-water heater. The hot water is then pumped through pipes to each room in the house. The hot water flows through heaters near the floor of each room. The hot water warms the air near the floor by conduction. The warm air rises, carrying heat through each room by means of convection currents. Heat circulates through each room in this way. After the water loses some of its heat, it returns to the hot-water heater to repeat the process.

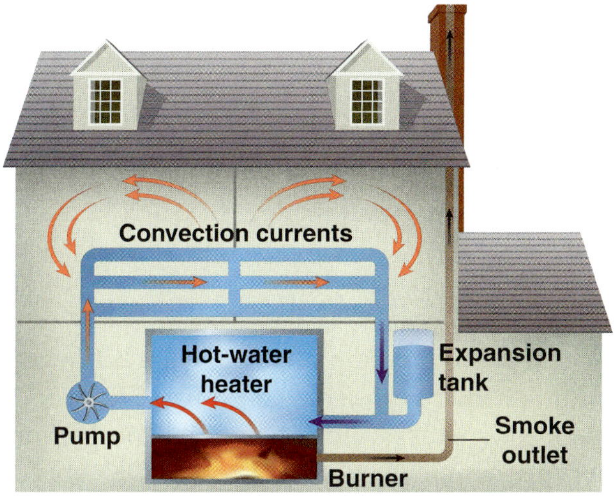

▲ **Figure 1-21** In a hot-water heating system, heated water travels through a system of pipes to heat a home.

3 IDENTIFY: What method of heat transfer is used in a hot-water heating system?

CHECKING CONCEPTS

1. Heat travels through gases by __________.
2. Warm air is __________ dense than cold air.
3. Warm water rises and cool water__________.
4. Convection takes place in gases and __________.

THINKING CRITICALLY

5. ANALYZE: To cool a room, should an air conditioner be placed at the bottom of a wall or at the top of a wall? Explain your answer.
6. INFER: Is heat transferred through solids by convection? Explain.

HEALTH AND SAFETY TIP

Smoke and gases from the burners of hot-water heaters contain pollutants. These pollutants are sent through an exhaust pipe that runs out through a chimney or vent in the roof. Hot air containing the pollutants rises up the chimney. Cool, clean air moves in from outside to take its place. Why do you think these pollutants need to be removed from a house?

Integrating Earth Science

TOPIC: plate tectonics

CONVECTION CURRENTS INSIDE THE MANTLE

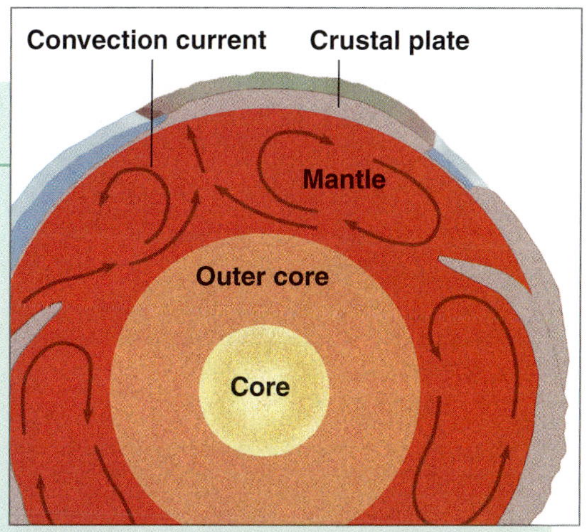

▲ **Figure 1-22** Convection currents in the mantle

Differences in density within a gas or liquid can cause convection currents within the gas or liquid. Under Earth's crust is a thick layer of rock called the mantle. The part of the mantle closest to Earth's core acts like a thick, hot liquid. The part of the mantle that is farthest from the core has the properties of a thick, cooler liquid. Scientists believe that the differences in the density in the mantle cause convection currents. The thick, hot liquid of the mantle rises and cools. The cooler liquid of the mantle sinks. As the cooler liquid gets closer to the core, it heats up and rises. The process repeats over and over again.

Scientists also believe that these convection currents cause the plates that make up the crust to continuously move. The crust floats on top of the mantle. The convection currents in the mantle carry the plates along so that the continents that sit on top of the plates slowly move, too. This concept supports the theory of plate tectonics. The theory of plate tectonics explains how Earth has changed over time.

Thinking Critically How do convection currents support the theory of plate tectonics?

LAB ACTIVITY
Creating Convection Currents

Materials
- Safety goggles
- Apron
- Deep storage container
- Small test tube
- Warm water
- Cold water
- Food coloring (red, blue)

▲ **STEP 3** Add red food coloring to the warm water in the test tube.

▲ **STEP 4** Gently lower the test tube to the bottom of the container.

BACKGROUND

How does an entire pot of soup on a stove get hot when only the bottom of the pot touches the stove? Heat travels through the metal pot to the soup by conduction. Then another process takes over. In this activity, you will study heat transfer by convection currents.

PURPOSE

In this activity, you will create convection currents in liquids.

PROCEDURE

1. Copy the chart shown in Figure 1-23 onto a sheet of paper. Put on safety goggles and an apron.

2. Fill the storage container almost to the top with cold water. If you need to move the box to your table, snap the lid on before carrying it.

3. Fill the test tube with very warm water almost to the top. Add a few drops of red food coloring to the water. Gently shake the test tube from side to side to mix the food coloring and water together.

4. Place your thumb or pointer finger over the mouth of the test tube. Very gently, lower the test tube into the storage container until the test tube reaches the bottom and settles over on its side.

5. Remove your finger from the mouth of the test tube. Observe what happens to the colored water.

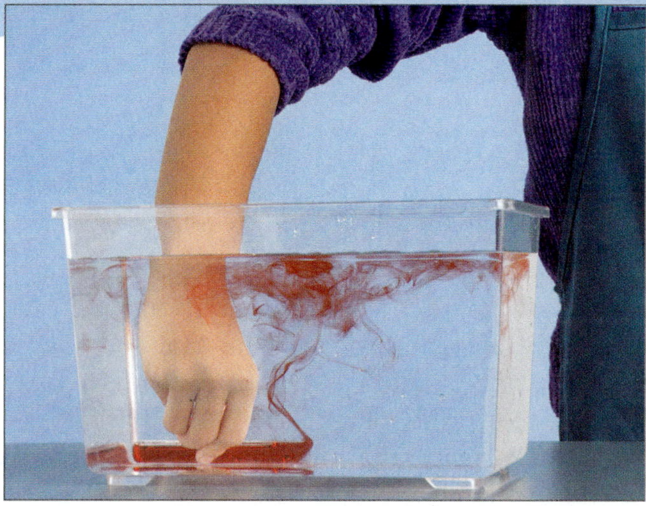

▲ **STEP 5** Observe what happens to the warm (red) water.

▲ **STEP 6** Remove your finger from the test tube near the surface of the water.

6. Repeat the experiment, this time using very warm water in the container and cold water, colored blue, in the test tube. Also, remove your finger from the test tube near the surface of the water. Observe what happens to the blue colored water.

Creating Convection Currents

Observations of container with cold water and test tube with warm water	*Observations of container with warm water and test tube with cold water*

▲ **Figure 1-23** Copy this chart and use it to record your observations.

CONCLUSIONS

1. **OBSERVE:** What happened to the red, warm water in the test tube when you placed it in the cold water?
2. **OBSERVE:** What happened to the blue, cold water in the test tube when you placed it in the warm water?
3. **ANALYZE:** What caused the colored water to move in both experiments?
4. **INFER:** What would happen if, in the first experiment, you held the test tube at the surface of the cold water?
5. **INFER:** What would happen in the second experiment if you lowered the test tube to the bottom of the box?

1-7 What is radiation?

Objective

Describe how heat travels through empty space.

Key Terms

vacuum: region where no matter exists

radiation (ray-dee-AY-shuhn)**:** transfer of energy through space

Heat from the Sun The Sun is about 150 million kilometers from Earth. Yet, Earth's surface is heated by energy from the Sun. In order for Earth to receive this energy, the energy must travel through the vacuum that exists in space between the Sun and Earth. You learned that a **vacuum** is a region where no matter exists. Because there are no particles of matter in a vacuum, the energy that reaches Earth cannot be transferred by conduction or convection.

1 DEFINE: What is a vacuum?

Heat Transfer Through Space Energy from the Sun is transferred to Earth by radiation. **Radiation** is the transfer of energy through space. Particles of matter are not needed for the transfer of heat energy by radiation. The heat is in the form of electromagnetic waves.

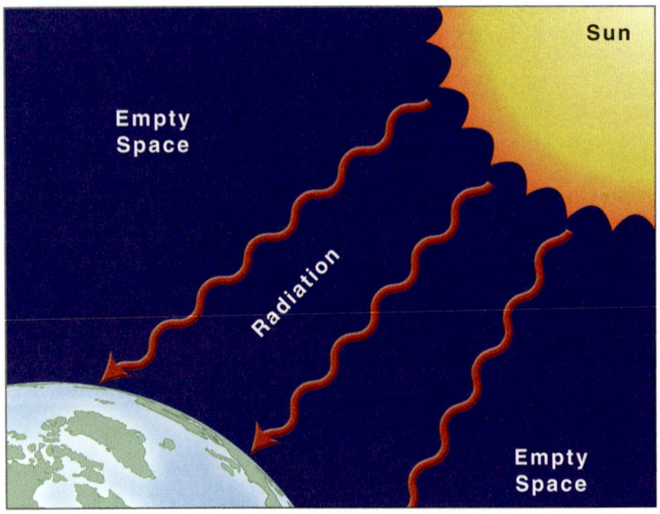

▲ **Figure 1-24** Radiation from the Sun travels through empty space.

2 IDENTIFY: What is radiation?

Another Source of Radiation The Sun is not the only source of radiation on Earth. Hot objects also radiate heat. For example, food is kept warm under hot lights in some restaurants, and an electric heater can heat a room. Have you ever toasted a marshmallow over an open fire? You do not have to put the marshmallow directly into the fire. If the marshmallow is close enough to the flame, heat radiates from the fire to the marshmallow and melts it.

▲ **Figure 1-25** Radiation from the fire heats the marshmallow.

3 IDENTIFY: List some sources of radiation.

Radiation in Home Heating Radiation from the Sun is used in some home heating systems. One type of heating system is called a passive solar-heating system. In this type of system, energy from the Sun heats a building directly. Large windows are usually placed where they will receive the maximum amount of sunlight. Energy from the Sun is absorbed by the walls and floors of the building and changed to heat. Shades covering the windows prevent heat loss at night.

Another type of heating system is an active solar-heating system. In an active solar-heating system, energy from the Sun is captured by solar collectors. Solar collectors are made of materials that absorb radiation from the Sun. They are usually installed on the roof or on the side of the house that faces the Sun. The energy captured by the solar collectors is used to heat water or air. The heated water or air is then circulated through the building. Because the amount of energy received from the Sun varies, backup heating systems are used in both passive and active solar-heating systems.

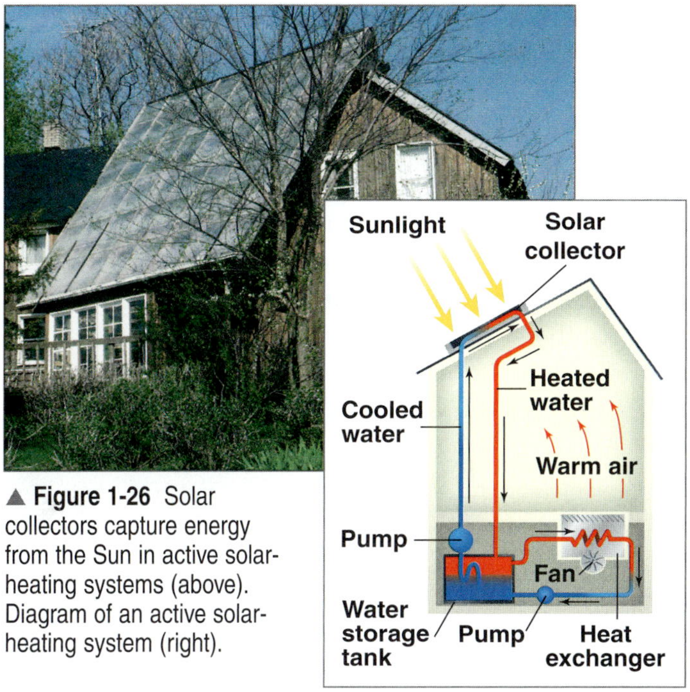

▲ **Figure 1-26** Solar collectors capture energy from the Sun in active solar-heating systems (above). Diagram of an active solar-heating system (right).

4 **NAME:** What are two kinds of heating systems that use radiation?

CHECKING CONCEPTS

1. Earth receives heat from the ___________.
2. Space where no matter exists is called a ___________.
3. Heat cannot travel from the Sun by ___________ or convection.
4. The transfer of energy through empty space is called ___________.
5. Two kinds of solar heating systems are active and ___________ solar heating.

THINKING CRITICALLY

6. **INFER:** Why does your face get warm when you turn it toward the Sun?

INTERPRETING VISUALS

Use the diagram in Figure 1-26 to answer the question.

7. **EXPLAIN:** How does the heat exchanger help create convection currents in the house?

Hands-On Activity

MEASURING THE EFFECT OF PASSIVE SOLAR HEATING

You will need two cardboard boxes, black construction paper, white paper, two thermometers, plastic wrap, scissors, and tape.

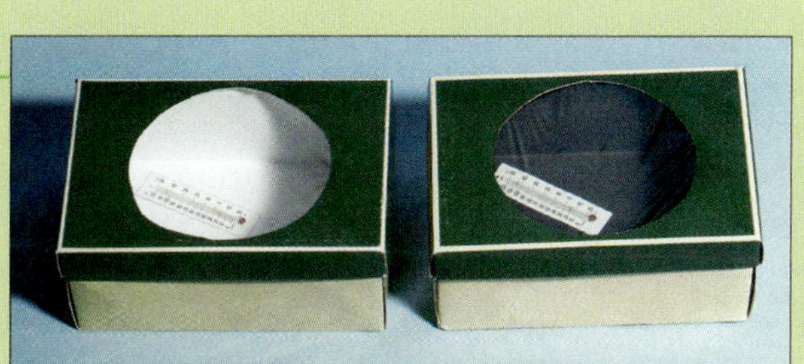

▲ **STEP 4** Place each box in direct sunlight.

1. Cut a large hole, or "window" in one side of each cardboard box. ⚠ CAUTION: Be careful when using scissors.
2. Completely line the inside of one box with black construction paper. Completely line the inside of the other box with white paper. Place a thermometer inside each box.
3. Cover each window with plastic wrap. Tape the plastic wrap tightly to the box.
4. Place each box in direct sunlight. Make sure that the window of each box fully faces the sunlight.
5. After about 20 minutes, open the boxes and read the temperature on each thermometer.

Practicing Your Skills

6. **MEASURE:** Which box got warmer?
7. **INFER:** Why did one box get warmer than the other?
8. **ANALYZE:** How does the box that got warmer model a passive solar heating system?

1-8 What is specific heat?

Objective

Describe the physical property of specific heat.

Key Term

specific heat: amount of heat needed to raise the temperature of 1 g of a substance 1°C.

Sand and Water Imagine a hot summer day at the beach. The sand is burning hot under your feet. You run into the water and it is "freezing" cold. How can the sand be so hot while the water is so cold? The answer is that different materials require a different amount of heat to raise their temperatures. Sand requires less heat to raise its temperature than water does. This is why the sand is so hot. However, water needs a much greater amount of heat to raise it to the same temperature as the sand. This is why the water is so much colder than the sand.

1 INFER: Which needs a greater amount of heat to raise its temperature, water or sand?

Specific Heat Adding heat to a substance increases its temperature. However, the rise in temperature for different types of substances is not the same, even if you have the same amount of each substance. For example, if you have equal amounts of sand and water, adding the same amount of heat to both samples will not give you the same temperature in both samples. The temperature of the sand will always be higher than the temperature of the water. The amount of heat needed to raise the temperature of any substance depends on the chemical makeup of the substance.

▲ **Figure 1-27** When heat is added to equal samples of sand and water, the temperature of the sand will always be higher than the temperature of the water.

The ability of a substance to absorb heat is called its specific heat. **Specific heat** is the amount of heat needed to raise the temperature of 1 g of a substance 1°C. For example, it takes 0.09 calories of heat to raise the temperature of 1 g of copper 1°C. Specific heat is a physical property. Every pure substance has its own specific heat. Figure 1-28 shows the specific heats of some substances.

SPECIFIC HEATS OF SOME SUBSTANCES	
Substance	Specific heat (cal /g°C)
Aluminum	0.22
Copper	0.09
Gold	0.03
Iron	0.11
Lead	0.03
Water	1.00

▲ **Figure 1-28**

2 INFER: If the same amount of heat is added to equal samples of gold and aluminum, which sample will have a higher temperature?

Applications of Low Specific Heat A substance with a low specific heat gets hot easily. Not only are metals such as copper and aluminum good conductors of heat, they also have low specific heats. This means that it does not take much heat energy to raise these metals to high temperatures.

3 PREDICT: Do substances with high specific heats get hot quickly or slowly?

Water Is Unique Water can take in or give off a lot of heat energy without a large change in temperature. Its specific heat is much higher than other earth materials, such as those that make up air and soil. That is why in winter, the air temperature around a lake can be as low as −20°C, but the water temperature may be 20°C. In summer, the air near the same lake can warm up to 25°C, whereas the water temperature remains a comfortable 12°C.

Water has an unusually high specific heat. One calorie of heat energy only raises the temperature of 1 g of water 1°C. Water's specific heat is about ten times greater than that of iron. This means that one calorie of heat raises the temperature of 1 g of iron about 10°C.

4 **HYPOTHESIZE:** On a very cold day, which is apt to be warmer, water in a lake or the air around the lake?

CHECKING CONCEPTS

1. Specific heat is the amount of heat needed to increase the temperature of __________ of a substance 1°C.
2. If a material has a __________ specific heat, it gets hot easily.
3. Iron gets __________ times hotter than water does when the same amount of heat is applied to both.
4. Specific heat is a __________ property.

THINKING CRITICALLY

5. **PREDICT:** Which would you expect to be hotter in the summer, a city near a large lake or a city that is far from any body of water? Explain your answer.
6. **ANALYZE:** A silver spoon and a stainless steel spoon are both placed in a cup of hot tea. The silver spoon becomes too hot to handle within minutes. The stainless steel spoon is warm. Explain the difference in the two substances.

DESIGNING AN EXPERIMENT

Design an experiment to solve the following problem. Include a hypothesis, variables, a procedure, and a type of data to study.

PROBLEM: Does ocean water have the same specific heat as pure water? If not, which has the higher specific heat?

Hands-On Activity

COMPARING THE SPECIFIC HEATS OF TWO METALS

You will need safety goggles, plastic gloves, an apron, two blocks of paraffin wax, a strip of copper, a strip of aluminum, a pencil, a beaker, water, a hot plate, tongs, and a timer.

1. Put on safety goggles, an apron, and plastic gloves.
2. Half-fill the beaker with water. Heat the beaker on a hot plate. ⚠ CAUTION: Hot plates and hot water can cause burns.
3. When the water boils, drop both metal strips into the beaker. Boil gently for 5 minutes.
4. Using the tongs, remove the copper strip from the beaker. Place it on top of one paraffin block.
5. Using the tongs, remove the aluminum strip from the beaker. Place it on top of the other paraffin block.
6. Observe the blocks of paraffin for 5 minutes.

▲ **STEP 6** Place the heated metal strips on blocks of paraffin wax. Observe for 5 minutes.

Practicing Your Skills

7. **IDENTIFY:** Which metal strip melted the most wax?
8. **INFER:** Which metal has the lower specific heat?

1-9 What is thermal expansion?

Objective

Describe what happens to solids, liquids, and gases when they are heated.

Key Term

thermal expansion: expansion of matter caused by heating

Heat and Expansion The expansion of matter caused by heating is called **thermal expansion.** Most materials expand, or get larger, when they are heated. You have learned that when heat energy is applied to a material, the particles of matter gain kinetic energy. The particles move faster and spread farther apart. As a result, the material expands. This happens even when the material does not change its state of matter.

What happens to a material as it cools? The particles move closer together. The material contracts, or gets smaller. Most solids contract when they are cooled.

1 **DEFINE:** What is thermal expansion?

Expansion in Solids Have you ever seen the spaces between the joints on a bridge? The spaces are there because they allow the metals in the bridge to expand. Metals, like most solids, expand when the weather gets hot. Without the spaces, the bridge could be badly damaged.

▲ **Figure 1-29** Spaces between the joints allow the metals in the bridge to expand.

Spaces are built into concrete sidewalks, railroad tracks, and other structures for the same reason. You may be familiar with expansion in solids, too. If you have ever placed the lid of a jar under hot water, you know that the heat from the water can make the lid expand enough to loosen it.

2 **EXPLAIN:** What happens to most solids when they are heated?

Expansion in Liquids Most liquids expand when they are heated. The particles of the liquid move farther apart as the liquid is heated. The opposite is usually true, too. When a liquid is cooled, the particles move closer together. The substance contracts, or gets smaller. However, water is an exception. Water contracts until it reaches a temperature of 4°C. Then it expands instead of contracting. The freezing point of water is 0°C. When water is cooled from 4°C to 0°C, it expands and its volume increases. As the volume of water increases, the density decreases. As a result, ice is less dense than water. This is why ice floats.

▲ **Figure 1-30** Ice floats because it is less dense than water.

3 **STATE:** What happens to water as it is cooled from 4°C to 0°C?

Expansion in Gases Gases also expand when they are heated and contract when they are cooled. Hot-air balloons rise because of the expansion of heated air.

Thermal expansion of a gas within a container can be dramatic. The particles in a gas are farther apart and move faster than the particles in a solid or a liquid. Adding heat to a gas will cause its particles to collide with one another and the sides of their container. If a gas in a container becomes too hot, the increased pressure can cause the container to burst. This is the reason why food should not be heated in a tightly sealed container.

4 **EXPLAIN:** How does the expansion of gas cause a hot-air balloon to rise?

CHECKING CONCEPTS

1. What is the expansion of materials caused by heating called?
2. What happens to most solids when they are cooled?
3. What happens to most liquids when they are heated?
4. What can happen if a gas in a container is heated?
5. What happens to gases when they are cooled?

THINKING CRITICALLY

6. **COMPARE:** What happens to the particles of most solids when they are heated? What happens to the particles in most solids when they are cooled?
7. **HYPOTHESIZE:** What would happen to a sidewalk on a hot day if there were no cracks between the squares in the sidewalk?

INTERPRETING VISUALS

Look at Figure 1-31 to answer the following question.

8. **ANALYZE:** Explain how the drawings illustrate thermal expansion.

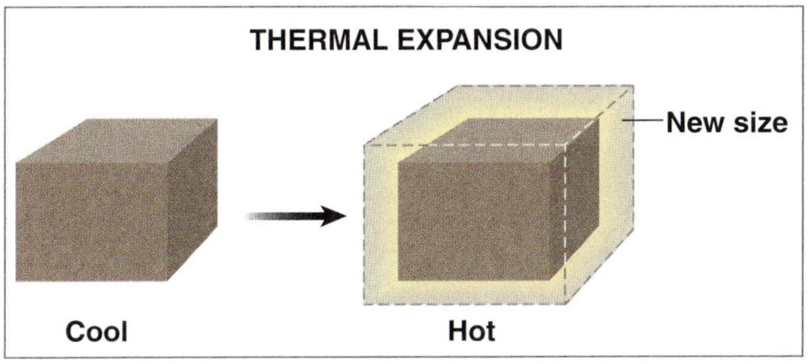

▲ **Figure 1-31** Thermal expansion

Science and Technology

BIMETALLIC THERMOSTATS

Thermal expansion can be used to help control how a home is heated and cooled. A device called a thermostat regulates the temperature in most homes. Inside the thermostat is a strip made of two metals. The device is called a bimetallic thermostat. The prefix *bi-* means "two."

The two metals in the thermostat expand when they are heated. However, they expand at different rates. One of the metals will expand faster than the other. It will also contract faster when cooled. As a result, the bimetallic strip bends when it is heated. It straightens out when it is cooled. This bending and unbending controls a switch.

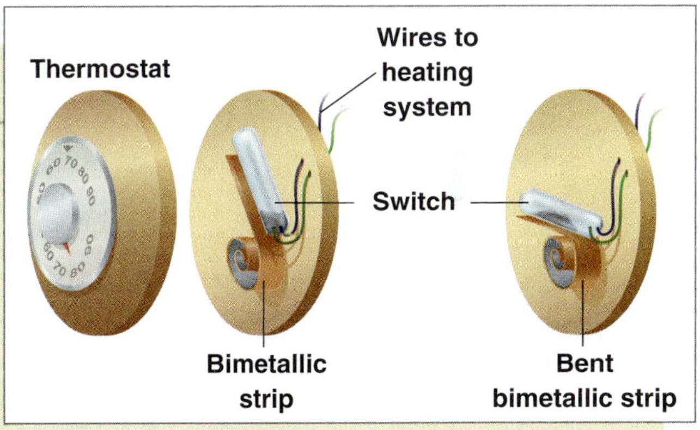

▲ **Figure 1-32** In a bimetallic thermostat, the bending and unbending of the bimetallic strip controls a switch.

The switch controls an electric circuit, which turns the home heating system on and off. For example, when the temperature in a room gets too high, the thermostat switches off the electric current. The heating system goes off, and the temperature drops. When the temperature gets too low, the thermostat switches the current back on. The heating system goes on, and the temperature rises.

Thinking Critically Why do you think the metal strip needs to be made of two different metals?

Integrating Life Science

THE Big IDEA

How do animals control their body temperatures?

Many animals are able to live in areas where the climate is extremely hot or extremely cold. The abilities of these animals to regulate their body temperatures and to behave in certain ways help them to live under extreme weather conditions.

Animals are classified into two groups—warm-blooded animals and cold-blooded animals. These terms are a bit confusing because both kinds of animals can be warm. Therefore, many scientists call warm-blooded animals endotherms (EHN-doh-thurms) and cold-blooded animals ectotherms (EHK-toh-thurms).

An endotherm regulates its body temperature to maintain a constant internal temperature. The endotherm has no control over this process. For example, when an endotherm is in cold surroundings, its body shivers to keep warm. When it is in hot surroundings, its body perspires or "sweats" to cool off.

The body temperature of an ectotherm can adjust to the temperature of the environment outside of its body. For example, the body temperature of an ectotherm that lives in water can decrease to better adapt to the temperature of the water. An ectotherm can also increase its body temperature by lying in the Sun. An ectotherm has some control over its body temperature because it can move from colder environments to warmer environments and vice versa.

Read about the endotherms and ectotherms that appear on these two pages. Then, follow the directions in the Science Log to find out more about "the big idea."✦

Walruses

Endotherms that live in cold environments have adaptations that help them to keep from losing heat. Sea mammals, such as these walruses, have a thick layer of insulating fat called blubber that retains body heat.

Lizards

A lizard is an ectotherm. It lies out in the Sun to absorb heat. Absorbing heat increases its body temperature.

Figure 1-33 Endotherms and ectotherms are found around the world.

Polar Bears

Polar bears are endotherms that maintain the same body temperature even when swimming in freezing water.

WRITING ACTIVITY

Science Log

On a map, look at the regions that are very cold or very hot. Pick a region and research what kinds of animals can be found there. In your science log, list the animals and classify them as either endotherms or ectotherms. Describe the behaviors that the animals have that help them live in their environments. Start your search at www.conceptsandchallenges.com.

Elephants

Some endotherms have to work to keep cool. Elephants enjoy baths to keep cool. Also, their large ears increase the surface area to radiate more heat out of their bodies.

Fishes

Ocean water can drop below freezing. The Arctic charr is an ectotherm that has a special protein in its blood that keeps its tissues from freezing.

Chapter 1 Challenges

Chapter Summary

Lesson 1-1

- **Heat** is energy of particles moving from warmer to cooler regions.
- All matter has heat.
- Heat energy moves from a warmer object to a cooler object.

Lesson 1-2

- **Temperature** is a measure of the average kinetic energy of the particles of a substance.
- Heat affects the temperature of a substance.

Lesson 1-3

- The temperature at which a liquid changes to a solid is its **freezing point.**
- The temperature at which a solid changes to a liquid is its **melting point.**

Lesson 1-4

- The temperature at which a liquid changes to a gas is its **boiling point.**

Lesson 1-5

- Heat moves through solids by **conduction.**
- Heat travels by conduction when moving particles of matter bump into one another.

Lesson 1-6

- Heat is transferred in gases and liquids by **convection.**
- **Convection currents** are found in liquids and gases.

Lesson 1-7

- Energy from the Sun travels through empty space.
- Transfer of energy through space is called **radiation.**

Lesson 1-8

- **Specific heat** is the amount of heat needed to raise the temperature of 1 g of a substance 1°C.

Lesson 1-9

- **Thermal expansion** is the expansion of a material caused by heating.

Key Term Challenges

boiling point (p. 22)
calorie (p. 18)
conduction (p. 24)
conductor (p. 24)
convection (p. 26)
convection current (p. 26)
evaporation (p. 22)
freezing point (p. 20)
heat (p. 16)
insulator (p. 24)
melting point (p. 20)
radiation (p. 30)
specific heat (p. 32)
temperature (p. 18)
thermal expansion (p. 34)
vacuum (p. 30)

MATCHING Write the Key Term from above that best matches each description.

1. change from a liquid to a gas at the surface of the liquid
2. total kinetic energy of all the particles in a sample of matter
3. region where no matter exists
4. transfer of heat through space
5. temperature at which a solid changes to a liquid
6. transfer of heat through a gas or liquid
7. measure of the average kinetic energy of the particles in a sample of matter
8. expansion of a material caused by heating

FILL IN Write the Key Term from above that best completes each statement.

9. The temperature at which a liquid changes to a solid is called the ___________.
10. The heat transfer in solids is called ___________.
11. Materials that do not conduct heat easily are called ___________.
12. The up and down movements of gases or liquids that cause heat to be transferred is called ___________.
13. A ___________ is a unit of heat.
14. Substances with low ___________ get hot easily.

Content Challenges TEST PREP

MULTIPLE CHOICE Write the letter of the term or phrase that best completes each statement.

1. Before Benjamin Thompson's experiment, people thought that heat was a
   - **a.** force.
   - **b.** gas.
   - **c.** form of energy.
   - **d.** substance.
2. A unit used to measure heat is the
   - **a.** calorie.
   - **b.** gram.
   - **c.** degree Celsius.
   - **d.** liter.
3. When water freezes, it changes to
   - **a.** heat.
   - **b.** ice.
   - **c.** caloric.
   - **d.** absolute zero.
4. Scientists now know that heat is a
   - **a.** force.
   - **b.** liquid.
   - **c.** form of energy.
   - **d.** substance.
5. When you add or remove heat from a material, you change its
   - **a.** mass.
   - **b.** weight.
   - **c.** potential energy.
   - **d.** temperature.
6. When water boils, it changes to
   - **a.** ice.
   - **b.** small particles.
   - **c.** a solid.
   - **d.** steam.
7. Water changes to ice at
   - **a.** 0°C.
   - **b.** 22°C.
   - **c.** 100°C.
   - **d.** 212°F.
8. Warm air
   - **a.** floats.
   - **b.** sinks.
   - **c.** rises.
   - **d.** turns into steam.
9. The freezing point of a substance is the same as its
   - **a.** boiling point.
   - **b.** elevation point.
   - **c.** dew point.
   - **d.** melting point.
10. Energy moves through a vacuum by
    - **a.** conduction.
    - **b.** convection.
    - **c.** radiation.
    - **d.** evaporation.

TRUE/FALSE Write *true* if the statement is true. If the statement is false, change the underlined term to make the statement true.

11. Adding heat energy makes particles of matter move slower.
12. Most solids expand as they freeze.
13. Absolute zero is about –273°C.
14. Particles of matter are always moving.

Concept Challenges TEST PREP

WRITTEN RESPONSE Complete the exercises and answer each of the following questions in complete sentences.

1. **COMPARE:** What is the difference between temperature and heat?
2. **EXPLAIN:** How is thermal expansion applied in a thermometer?
3. **HYPOTHESIZE:** Do you think it is possible to reach absolute zero in a laboratory? Why or why not?
4. **COMPARE:** Describe the differences between the transfer of heat by conduction, convection, and radiation.
5. **EXPLAIN:** How did Benjamin Thompson's experiment disprove the caloric theory?

INTERPRETING A TABLE Use Figure 1-34 to answer the following questions.

6. Which substance has the highest specific heat?
7. Which substance has the lowest specific heat?
8. Which substance has a specific heat of 0.22 cal/g °C?
9. Which substance heats up faster, copper or iron?
10. Which substance heats up slower, gold or aluminum?

SPECIFIC HEATS OF SOME SUBSTANCES	
Substance	**Specific heat (cal /g°C)**
Aluminum	0.22
Copper	0.09
Gold	0.03
Iron	0.11
Lead	0.03
Water	1.00

▲ **Figure 1-34**

Chapter 2 Waves

▲ **Figure 2-1** A pebble dropped in a pond creates waves in a circular pattern.

A pebble falls and disturbs the smooth, calm surface of a pond. The energy of the falling pebble is transferred to the water. Small waves called ripples move out in all directions from the place where the pebble entered the water. These waves carry the energy to the edges of the pond.

▶ What will happen when the water drops from the splash fall back to the pond's surface?

Contents

2-1 What is a wave?

INVESTIGATE

Observing How Waves Travel
HANDS-ON ACTIVITY

1. Fill a shallow pan with water to a depth of 3 cm.
2. Using a pencil or ruler, repeatedly touch the surface of the water at one end of the pan for about 1 minute. Sketch what you see happening to the water's surface.
3. Place a cork in the middle of the pan. Predict how it will move with the waves.
4. Repeat Step 2 to test your prediction.

THINK ABOUT IT: What does the motion of the cork tell you about the movement of waves?

Objective
Identify a wave as energy traveling through a medium.

Key Terms
wave: disturbance that transfers energy from place to place

mechanical wave: wave that transfers energy through matter

electromagnetic wave (ee-lehk-troh-mag-NEHT-ihk)**:** wave that transfers energy through empty space

medium: material through which mechanical waves can travel

Waves and Energy If you have ever stood in heavy surf, like that shown in Figure 2-2, you know that water waves have a lot of energy. **Waves** are disturbances that transfer energy from place to place.

▼ **Figure 2-2** There is a great amount of energy in heavy surf.

Mechanical waves transfer energy through some form of matter. **Electromagnetic waves,** such as light and radio waves, do not need matter. They can transfer energy through empty space. You will learn more about this type of wave in a later chapter.

▲ **Figure 2-3** The swimmer and his feathered friends create V-shaped waves as they move through the water. These waves transfer energy supplied by the swimmers.

1 **DEFINE:** What are waves?

Mechanical Waves Mechanical waves can travel only through some kinds of matter. Any material through which mechanical waves travel is called a **medium.** The picture in Figure 2-3 shows how water can serve as a medium. Air serves as a medium for sound waves.

When a wave travels through a medium, only energy moves from place to place. The particles of the medium do not move forward with the wave. Think of a bottle floating on water as shown in Figure 2-4. What happens as a wave moves past the bottle? The bottle moves up and down. It does not move forward in the same direction as the wave. The wave moves through the water.

2 **STATE:** Do all mechanical waves need a medium?

Particle Motion in a Medium When a wave travels through a medium such as water, the particles of the medium do not move forward with the wave. Figure 2-4 shows a wave moving through water. As described earlier, the bottle does not travel with the wave. It bobs up and down. This up-and-down motion of the bottle is caused by the motion of the water particles.

As a wave moves through the water, each water particle follows a circular path. The size of the circles is greatest at the surface of the water and decreases with depth. The water does not move forward with the wave. The energy of the wave is transferred forward to the next group of water particles, which also move in circles.

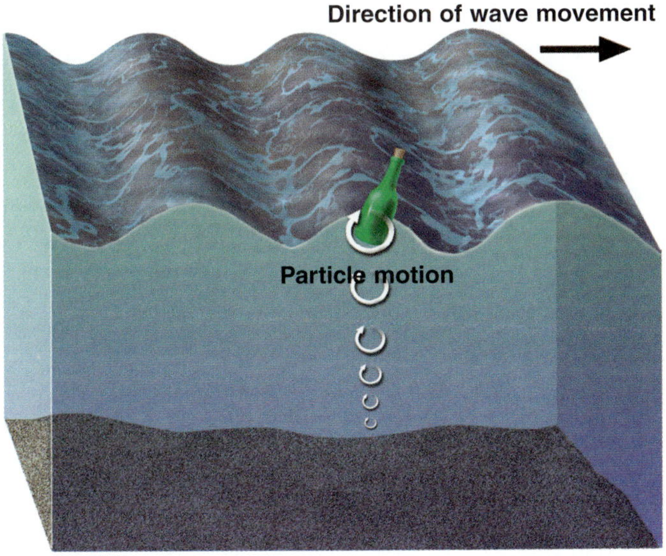

▲ **Figure 2-4** As a wave passes, a bottle will bob up and down but will not move forward.

3 **DESCRIBE:** What happens to water particles as a wave moves through the water?

CHECKING CONCEPTS

1. Water waves carry ___________.
2. Water is a ___________ for waves.
3. Particles move in ___________ as wave energy moves forward.
4. Some waves, such as ___________ waves, do not require a medium.
5. Light waves can travel through ___________.
6. When a wave moves through a medium, only ___________ moves from place to place.

THINKING CRITICALLY

7. **EXPLAIN:** You are sitting in a rowboat in the middle of a lake when a motorboat passes by, making waves that hit your boat. Describe what happens to your boat.
8. **ANALYZE:** Does the engine of a space shuttle produce sound as it orbits Earth?

HEALTH AND SAFETY TIP

During a hurricane or other large storm, a great deal of energy is carried by ocean waves. Storm waves can cause serious damage when they hit the shore. You should stay away from the shore during a hurricane. Look at Figure 2-5 for a safety checklist. Check with your local Red Cross for anything specific in your area.

HURRICANE SAFETY TIPS	
What You Need	**What You Have to Do**
☐ A full gas tank in your car	☐ Tape windows
☐ Necessary medications	☐ Close shutters
☐ First-aid kit	☐ Stay inside or evacuate
☐ Canned goods	☐ Fill tub with water
☐ Fresh batteries	☐ Tie down movable items outside
☐ Flashlight	☐ Stay on the side of the house away from the wind
☐ Portable radio	☐ Beware of downed power lines and broken gas mains

▲ **Figure 2-5**

2-2 How do waves travel through matter?

INVESTIGATE

Observing Waves in a Rope
HANDS-ON ACTIVITY

1. Tie a colored ribbon to the middle of a 3-m length of rope.
2. Tie one end of the rope to a doorknob. Hold the other end of the rope and stand opposite the door.
3. Quickly move your end of the rope up and down. Observe the ribbon's motion.
4. Increase the speed at which you move the end of the rope up and down. Observe the resulting waves.

THINK ABOUT IT: a. What happened to the ribbon when you moved the rope? **b.** What happened when you increased the speed of your movements? **c.** What happened to the wavelength?

Objective
Classify waves as transverse or longitudinal.

Key Terms
transverse (trans-VUHRS) **wave:** wave in which the particles of the medium move up and down at right angles to the direction of the wave motion

crest: high point of a transverse wave

trough (TRAWF)**:** low point of a transverse wave

longitudinal (lahn-juh-TOOD-uhn-uhl) **wave:** wave in which the particles of the medium move back and forth in the direction of the wave motion

compression (kuhm-PRESH-uhn)**:** part of a medium where the particles are close together

rarefaction (rer-uh-FAK-shuhn)**:** part of a medium where the particles are far apart

Transverse Waves There are two types of mechanical waves, transverse and longitudinal. The difference between the two kinds of waves is the way the particles of the medium move. In a **transverse wave,** the particles of the medium move at right angles, or perpendicular, to the direction of the wave motion. Water waves are transverse waves.

There are two parts to a transverse wave. The **crest** is the high point of a transverse wave. The **trough** is the low point of the wave.

You can demonstrate a transverse wave by tying one end of a rope to a doorknob and jerking the other end with a sharp up-and-down motion as shown in Figure 2-6.

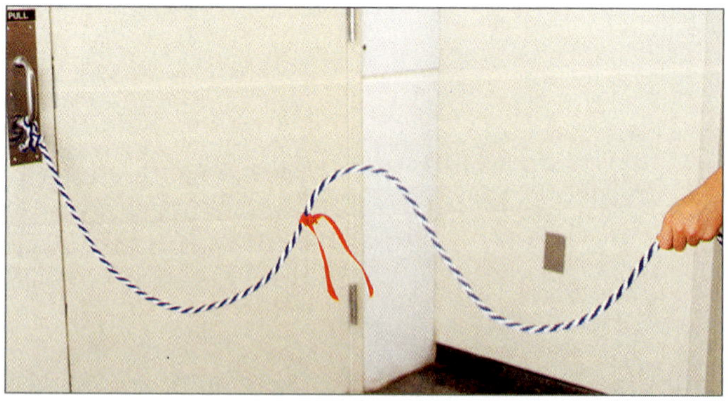

▲ **Figure 2-6** A model of a transverse wave

1 LIST: What are the two parts of a transverse wave?

Longitudinal Waves The sound of thunder moves in a series of longitudinal waves. A **longitudinal wave** is a wave in which particles of the medium move back and forth, parallel to the direction of the wave motion. The air is the medium that carries the energy of the thunder clap.

A longitudinal wave has two parts. A clap of thunder pushes the particles of air close together. This part of the wave is called a **compression.** The compressed particles move forward in the direction of the wave motion. As the particles move forward, they leave behind part of the wave where the particles are far apart. This part of the wave is called a **rarefaction.** The rarefaction also moves forward.

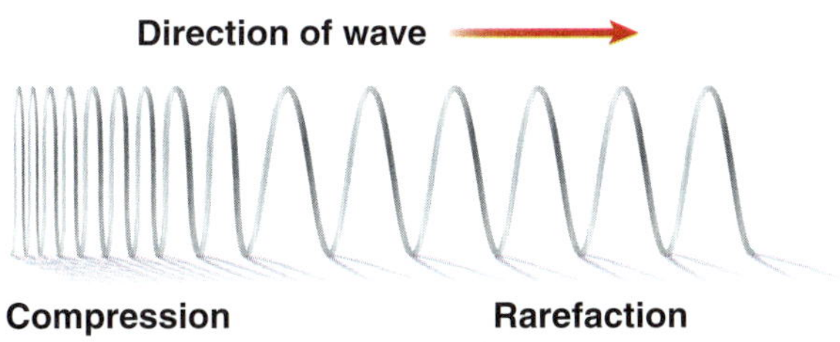

▲ **Figure 2-7** A model of a longitudinal wave

2 **DEFINE:** What are rarefactions?

CHECKING CONCEPTS

1. All waves carry ____________.
2. The difference between waves depends on how the ____________ of the medium move.
3. The particles of the medium move up and down in a ____________ wave.
4. The particles of the medium move back and forth in a ____________ wave.
5. The parts of a transverse wave are the crest and the ____________.

THINKING CRITICALLY

6. **INFER:** Are ocean waves transverse waves or longitudinal waves? How do you know?
7. **CLASSIFY:** Clap your hands together. What kind of wave did you make?
8. **INFER:** Have you ever seen fans do "the wave" at a baseball or football game? What kind of wave did they make?

Integrating Earth Science

TOPICS: earthquakes, seismic waves

EARTHQUAKE WAVES

In October 1989, a large earthquake struck San Francisco and Oakland, California. The earthquake registered 7.1 on the Richter scale. The Richter scale is a measure of how much energy an earthquake releases. The California earthquake of 1989 was very powerful.

The energy of an earthquake produces waves that travel through the Earth. All earthquakes produce three main types of waves. These waves are called primary waves (P-waves), secondary waves (S-waves), and long waves (L-waves). L-waves are also called surface waves. They cause the surface of the Earth to rise and fall. L-waves cause the greatest damage during an earthquake.

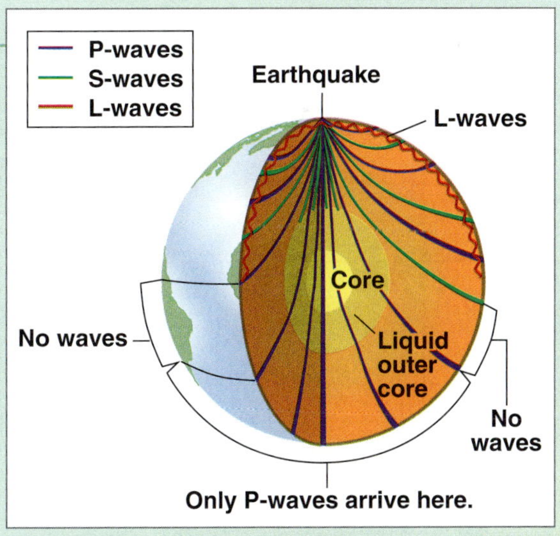

▲ **Figure 2-8** A diagram of three kinds of seismic waves

Scientists record earthquake waves on an instrument called a seismograph. P-waves, S-waves, and L-waves travel at different speeds. Scientists calculate the difference in arrival times of the three waves. They use this information to locate the area of Earth's surface directly above the earthquake's origin.

Thinking Critically Which waves cause the most damage? Explain your answer.

2-3 What are the features of a wave?

Objectives

Describe the features of a wave. Relate wave speed, frequency, and wavelength.

Key Terms

wavelength: distance between two neighboring crests or troughs

frequency (FREE-kwuhn-see)**:** number of complete waves passing a point in a given time

amplitude (AM-pluh-tood)**:** height of a transverse wave

speed: distance a wave travels in one unit of time

hertz (HURTS)**:** unit used to measure the frequency of a wave

Features of Transverse Waves All waves have some basic features. These features are wavelength, frequency, and amplitude. Figure 2-9 shows these features in a transverse wave.

Rest position

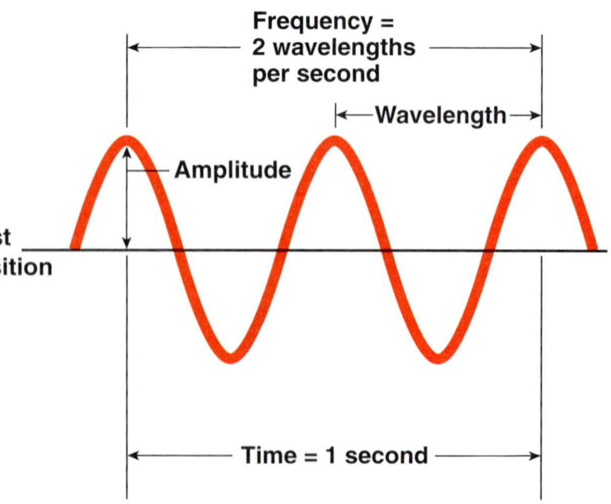

▲ **Figure 2-9** Features of a transverse wave

All waves have a certain length. The distance from the crest or trough of one wave to the crest or trough of the next wave is the **wavelength.** Wavelength can be measured in meters or centimeters.

A certain number of waves pass a point in a given amount of time. The number of complete waves per unit of time is called **frequency.** Frequency is expressed in waves per second.

When a wave moves through a medium, the particles of the medium are moved from their rest position. The height the particles are moved is called the **amplitude** of the wave.

1 LIST: What are some basic features of a wave?

Features of Longitudinal Waves Longitudinal waves have the same features as transverse waves. The distance from one compression to another or from one rarefaction to another is the wavelength. The number of waves that pass a certain point each second is the frequency. The amplitude of a longitudinal wave depends on the amount of energy in the wave. This is shown by how tightly or loosely compressed the particles are.

2 DEFINE: What is the amplitude of a longitudinal wave?

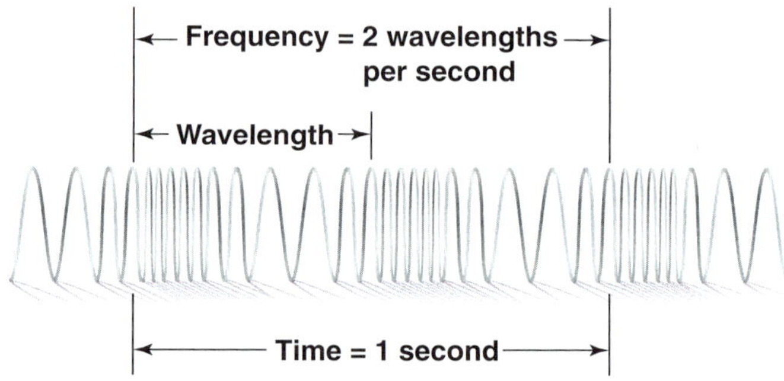

▲ **Figure 2-10** Features of a longitudinal wave

Speed of a Wave All waves move. Their speed of movement is related to their frequency and wavelength. **Speed** is the distance a wave travels in one unit of time. Wave speed is equal to frequency times wavelength.

$$\text{speed} = \text{frequency} \times \text{wavelength}$$

Scientists use a unit called a **hertz** to measure frequency. One hertz (Hz) is equal to one wave per second. When frequency is measured in hertz and wavelength is measured in meters, speed is measured in meters per second (m/s).

3 IDENTIFY: What is the equation used to find the speed of a wave?

CHECKING CONCEPTS

1. All waves have amplitude, __________, and frequency.
2. Amplitude is the __________ of a transverse wave.
3. Wavelength can be measured in __________ or centimeters.
4. The number of __________ passing a point in a given time is called frequency.
5. The speed of a wave is equal to __________ multiplied by wavelength.
6. The __________ is the unit used to measure frequency.
7. One hertz is equal to one __________ per second.

THINKING CRITICALLY

8. **LIST:** Name the parts of a transverse wave and the corresponding parts of a longitudinal wave.
9. **COMPARE:** What is the difference between amplitude in a transverse wave and amplitude in a longitudinal wave?

BUILDING MATHEMATICS SKILLS

Use the equation to complete the following questions:

speed = frequency × wavelength

a. A wave has a frequency of 50 Hz and a wavelength of 10 m. What is the speed of the wave?

b. The speed of a wave is 5 m/s. Its wavelength is 2 m. What is the frequency of the wave?

c. The frequency of a wave is 20 Hz. Its speed is 100 m/s. What is the wavelength of the wave?

How Do They Know That?

HEINRICH HERTZ

▲ **Figure 2-11** Heinrich Hertz

During the 1800s, the physicist James Clerk Maxwell developed a theory of electromagnetism. Using Maxwell's findings, Henrich Hertz, a physics professor, sought to perfect Maxwell's theory. Hertz conducted a series of experiments in which he proved that electricity can be transmitted in electromagnetic waves. He set up a condenser using two metal rods. He placed the rods end to end, leaving space between them small enough for a spark. When charged, the rods produced a spark that traveled back and forth between them. Thus, Hertz proved that the velocity of radio waves can be timed. Their velocity is the same as that of light. He was also able to solve the problem of how to free the waves from electric and magnetic fields.

A young inventor used Hertz's findings to develop both the telegraph and the radio. Radar is also based on Hertz's work.

The unit used to measure frequency of waves, Hertz (Hz), is named after this great physicist.

Thinking Critically What important discovery concerning electricity was made by Hertz?

LAB ACTIVITY
Making Waves

Materials
- Safety goggles
- Coil spring
- Meter stick
- Clock with second hand
- Stopwatch

BACKGROUND

One of the ways energy is transmitted is through waves. Water waves carry kinetic energy across oceans to pound on distant shorelines. Sound is carried in waves that cause the eardrums in our ears to vibrate. How do waves carry energy?

PURPOSE

You will create waves in a spring and observe their properties.

PROCEDURE

1. Select an area in which to conduct your experiment. The test area should be 3 to 4 meters long.

2. Copy the chart in Figure 2-12 onto a sheet of paper. Put on safety goggles.

3. Select two members of your group to operate the springs for the first tests.

4. Stretch the spring the length of the test area.

5. Begin making waves in the spring by moving just one end of the spring side to side. The movement should be slow and easy.

6. Measure the length of the waves produced. Then, use the stopwatch to count how many waves arrive at the other end of the spring each second (frequency). Record your observations.

7. Increase the number of waves by moving one end of the spring faster. Again, measure the wavelength and the number of waves arriving at the other end each second.

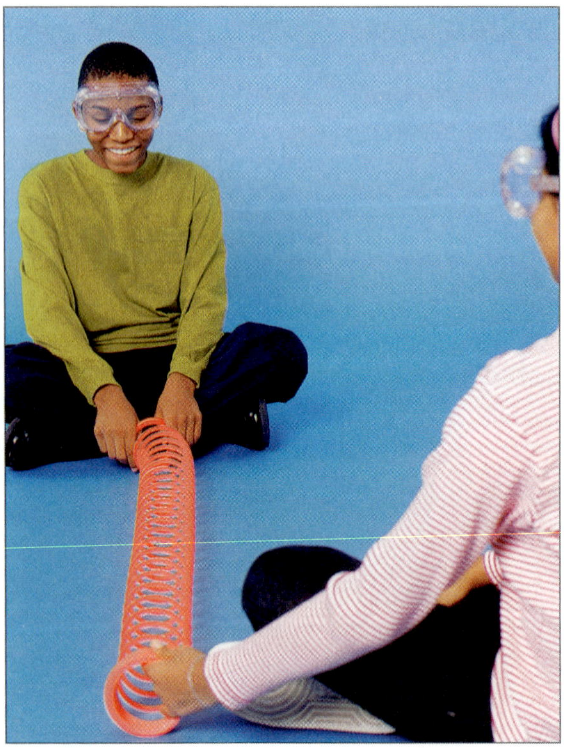

▲ **STEP 4** Stretch the spring.

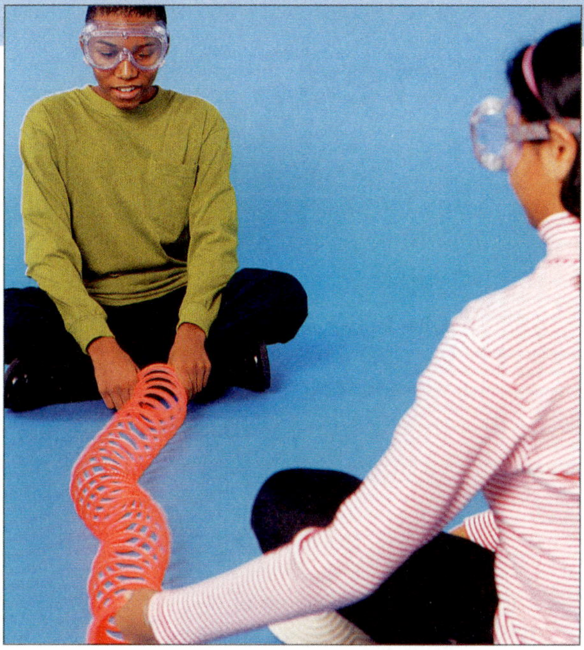

▲ **STEP 5** Move one end side to side to make waves.

▲ **STEP 7** Move the spring faster to increase the wave frequency.

8. Increase the number of waves again and make the same observations as before.
9. Switch jobs in your group and repeat the tests.

Wavelengths, Frequency, and Energy

	Wavelength (length between crests)	Frequency (waves per second)	Energy Needed (small, medium, or large)
Test 1			
Test 2			
Test 3			

▲ **Figure 2-12** Copy this chart and use it to record your observations.

CONCLUSIONS

1. **OBSERVE:** When the wavelengths are long, is the frequency low or high?
2. **OBSERVE:** When the wavelengths are short, is the frequency low or high?
3. **ANALYZE:** Which waves are easier to make, short waves or long waves?
4. **INFER:** If you hear a low sound, what can you infer about its frequency?
5. **INFER:** If you hear a high sound, does it have a low frequency or a high frequency?

2-4 How are waves reflected?

Objectives

Describe what happens when a wave strikes a barrier. State the law of reflection.

Key Terms

reflection: bouncing of a wave after striking a barrier

incident wave: wave that strikes a barrier

reflected wave: wave that bounces off a barrier

normal: line at right angles to a barrier

Waves and Barriers What happens when a wave hits a barrier like the rocks in Figure 2-13? Remember that all waves carry energy. Some of the wave's energy may be absorbed by the barrier. If the barrier does not absorb the wave's energy, the wave bounces off the barrier. This bouncing of a wave from a barrier is called **reflection.**

▲ **Figure 2-13** An incident wave strikes a barrier and is reflected.

1 DESCRIBE: What happens when a wave strikes a barrier that does not absorb all of its energy?

Reflection Some familiar examples of reflections include a ball bouncing off a wall, your image coming from a mirror, and an echo in an empty room. Figure 2-14 shows what happens when a wave strikes a barrier. The red arrows show the direction of the wave. The wave that strikes the barrier is called the **incident wave.** The wave that bounces off the barrier is called the **reflected wave.**

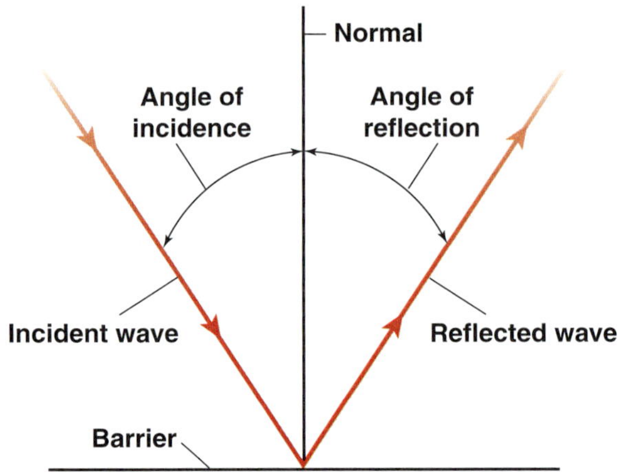

▲ **Figure 2-14** The angle of incidence equals the angle of reflection.

2 DEFINE: What is a reflected wave?

Law of Reflection The law of reflection describes what happens when a wave is reflected from a barrier. The angle at which an incident wave strikes a barrier is called the angle of incidence, or *i*. The angle at which the wave is reflected is called the angle of reflection, or *r*. These angles are measured from a line called the normal. The **normal** is a line at a right angle to the barrier. A right angle is equal to 90°. The law of reflection states that the angle of incidence is equal to the angle of reflection. Suppose a wave strikes a barrier at a 45° angle. The reflected wave will bounce off the barrier at a 45° angle.

3 STATE: What is the law of reflection?

CHECKING CONCEPTS

1. What happens when a wave strikes a barrier?
2. What is reflection?
3 What is a wave that strikes a barrier called?
4. What is a wave that bounces back from a barrier called?
5. What is the normal?
6. What is the angle formed by the normal and the barrier?

Web InfoSearch

Standing Waves Use the Internet to find out what standing waves are and how they are formed. Write a report of your findings. Include a diagram of standing waves in a rope. Start your search at www.conceptsandchallenges.com. Some key search words are **standing waves** and **reflected waves.**

THINKING CRITICALLY

Use Figure 2-15 to answer the following questions.

7. CONTRAST: What is the difference between the angle of incidence and the angle of reflection?
8. ANALYZE: Which arrow represents the incident wave?
9. ANALYZE: Which arrow represents the reflected wave?
10. ANALYZE: Which angle is the angle of incidence?
11. ANALYZE: Which angle is the angle of reflection?

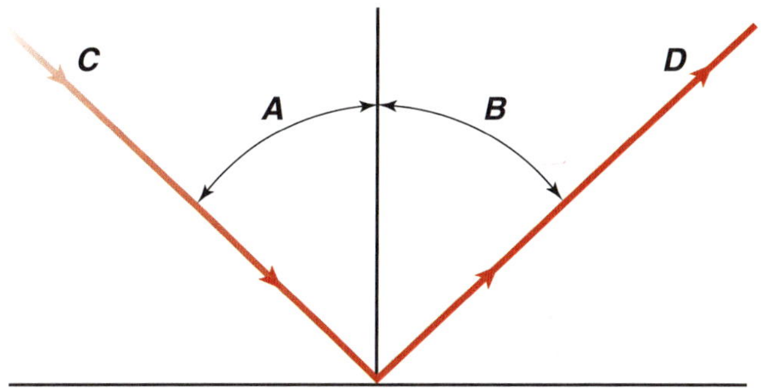

▲ Figure 2-15

Hands-On Activity

MEASURING THE ANGLE OF INCIDENCE AND THE ANGLE OF REFLECTION

You will need a small rubber ball, dark construction paper, a protractor, ruler, a small amount of talcum powder, and a marking pen.

1. Fold the construction paper in half lengthwise. Open the paper and lay it on the floor with the short side against the wall. The line of the fold represents the normal.
2. Measure an angle of 45° between the wall and the normal and mark it. Spread a small amount of talcum powder on this side of the normal.
3. Lay a ruler along the marked angle as shown. Slowly roll the ball along the edge of the ruler.
4. Observe the ball as it bounces off the wall. It should leave a light path of powder. Mark the angle of this path from the wall and measure it.
5. Label the angles of incidence and reflection.

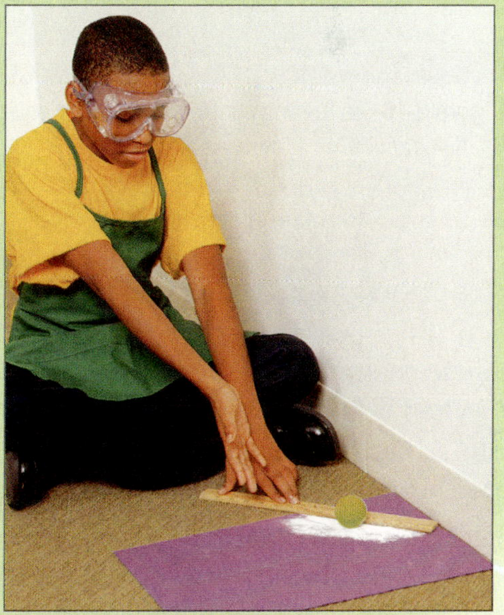

▲ **STEP 3** Using the ruler as a guide, roll the ball toward the wall.

Practicing Your Skills

6. MEASURE: What is the size of the angle of incidence?
7. MEASURE: What is the size of the angle of reflection?

2-5 How are waves refracted?

Objective

Describe what happens to a wave when it moves from one medium to another.

Key Term

refraction: bending of a wave as it moves from one medium to another

Changing the Medium Waves travel in straight lines through a medium. What happens to a wave when it moves from one medium to another? Suppose a wave moves from air into water. If the wave enters the water at an angle other than 90°, the wave bends. This bending of a wave as it moves from one medium to another is called **refraction.**

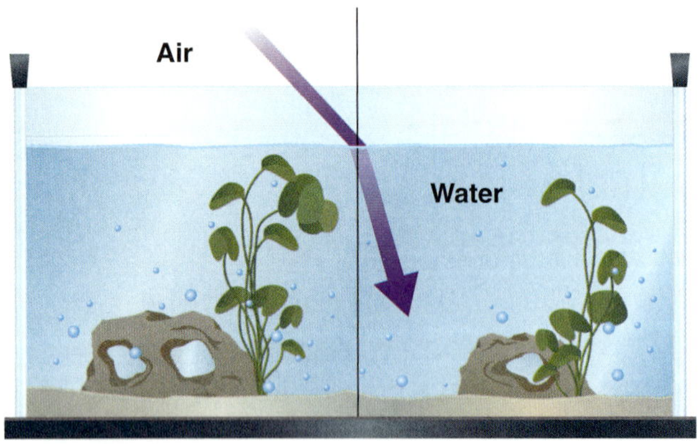

▲ **Figure 2-16** As the wave leaves the air and enters the water, it is refracted.

1 DEFINE: What is refraction?

Refraction and Wave Speed Waves bend when they go from one medium to another because they change speed. Water is denser than air. When a light wave moves from air into water, it slows down. When a light wave moves from water into air, it speeds up. This change in speed causes the wave to be refracted, or bent.

You can see the results of refraction by performing a simple experiment. Place a pencil into a glass of water at an angle. The pencil appears to be broken where it enters the water. As light waves move from air into water, they slow down. This change in speed causes the light waves to be refracted. As a result, the pencil appears broken.

◀ **Figure 2-17** A pencil placed in water at an angle appears broken.

2 EXPLAIN: What causes refraction as waves move from one medium to another?

Laws of Refraction The three laws of refraction describe how waves are refracted when they move from one medium to another.

- When a wave moves at an angle from a less dense medium to a more dense medium, it is bent toward the normal.
- When a wave moves at an angle from a more dense medium to a less dense medium, it is bent away from the normal.
- When a wave moves from one medium to another along the normal, it is not bent.

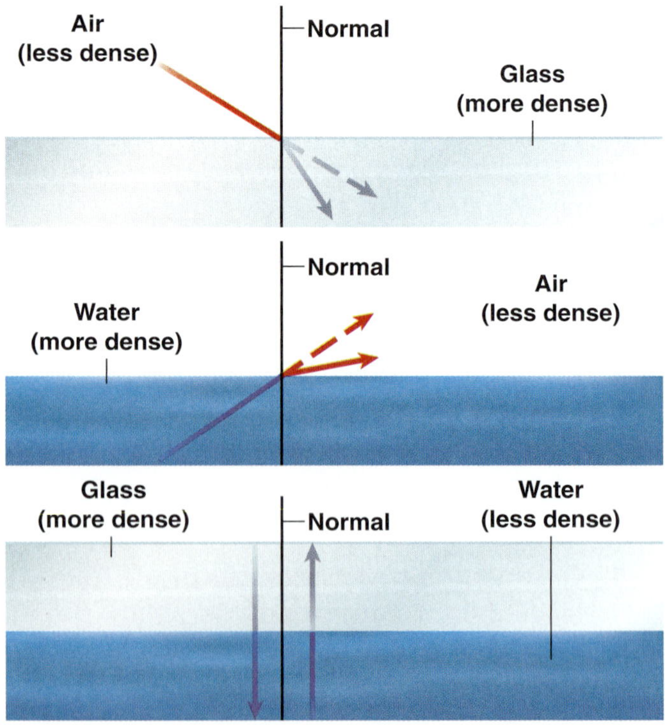

▲ **Figure 2-18** These diagrams illustrate the three laws of refraction.

3 DESCRIBE: In what direction is a wave bent when it moves at an angle from a more dense medium to a less dense medium?

CHECKING CONCEPTS

1. Waves travel through a medium in __________ lines.
2. When a wave moves at an angle from one medium to another, it __________.
3. The bending of a wave is called __________.
4. Refraction is caused by a change in __________.
5. The speed of a light wave __________ when it moves from water into air.
6. When a wave moves at an angle from a less dense medium to a more dense medium, it is bent __________ the normal.

THINKING CRITICALLY

Use Figure 2-19 to answer question 7.

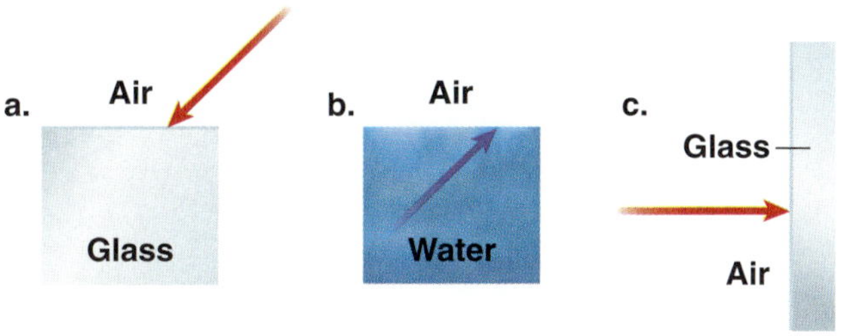

▲ Figure 2-19

7. **ANALYZE:** Copy the diagrams onto a separate sheet of paper. Draw arrows to show how the light waves will be refracted in each example.
8. **HYPOTHESIZE:** Have you ever tried to pick up an object underwater? Why do you think the object appeared closer to the surface of the water than it really was?

Web InfoSearch

Ocean Waves Ocean waves change direction as they come close to a shoreline. Waves almost always approach the shore at an angle. However, they usually hit the shore straight on. As waves approach the shallow water near shore, they slow down. This change in speed causes the waves to change direction.

SEARCH: Use the Internet to find out more about the direction of ocean waves. Start your search at www.conceptsandchallenges.com. Some key search words are **ocean waves, wave,** and **ocean wave refraction.**

Hands-On Activity

OBSERVING THE EFFECTS OF REFRACTION

You will need safety goggles, a small bowl, water, and a penny.

1. Place a penny into a small bowl.
2. Move away from the bowl until you can no longer see the penny.
3. Without changing your position, have a partner add water to the bowl until you can see the penny again.

▲ **STEP 3** Add water until your partner can see the penny.

Practicing Your Skills

4. **ANALYZE:** Why could you not see the penny when you moved away from the bowl?
5. **HYPOTHESIZE:** Why could you see the penny again after water was added to the bowl?
6. **MODEL:** Draw a diagram showing how light waves reflected from the penny were refracted as they moved from the water into the air.

Integrating Earth Science

THE Big IDEA

Tsunami: How big can a wave get?

How big can an ocean wave get? Think about a wave 18 m taller than average that is traveling at 800 km/h (480 mph). Does it sound like a horror movie? Such waves actually exist. They are called seismic sea waves or *tsunamis*, a Japanese word meaning "harbor wave." They occur most commonly in the Pacific.

Tsunamis are fast, high-energy waves related to earthquakes or volcanic eruptions on the seafloor. Sometimes people call them tidal waves. However, they are not caused by tides. A tsunami is caused by seismic activity, which is the movement of the Earth's crust. Energy from violent seismic movement on the ocean floor is transmitted into the water as a waveform. The high-energy wave can travel thousands of kilometers from its source.

In the deep open ocean, a tsunami might be only 1 m high, but when it reaches shore, watch out. The wave crest might be 18 m higher than normal waves.

The two worst tsunamis in this century struck Hawaii in 1946 and 1960. The port of Hilo on the Big Island of Hawaii was hit the worst. The 1960 tsunami killed 61 people.

Now, Hawaii has a tsunami warning system to protect people. When seismographs (earthquake detectors) detect a seafloor earthquake starting a tsunami, loud alarms go off along the coast warning people to head for the hills. Surprisingly, it is also safe to be out at sea. In the open ocean, a passing tsunami would just feel like a large wave. Its awesome power is only revealed when the tsunami moves into the shallow water near the shore.

Look at the illustrations and photos that appear on these two pages. Then, follow the directions in the Science Log to find out more about "the big idea."✦

Figure 2-20 Woodcut by Japanese artist Hokusai

Tsunami Damage

Out at sea a tsunami doesn't appear much larger than other waves. However, when the high energy of the tsunami hits the shore, it can be very destructive. In 1960, the port of Hilo in Hawaii suffered great damage.

Science Log

WRITING ACTIVITY

Find out how the tsunami warning system in Hawaii works. Your job is running part of the warning system. What data do you need to collect? How will you analyze the data? What do you think are the best ways to warn people? How will you make sure the people of your island are prepared to respond to a potential tsunami? To research a warning system, visit www.conceptsandchallenges.com.

Seismograph

Seismographs detect the seafloor earthquakes that start a tsunami. That is the key to the early warning system.

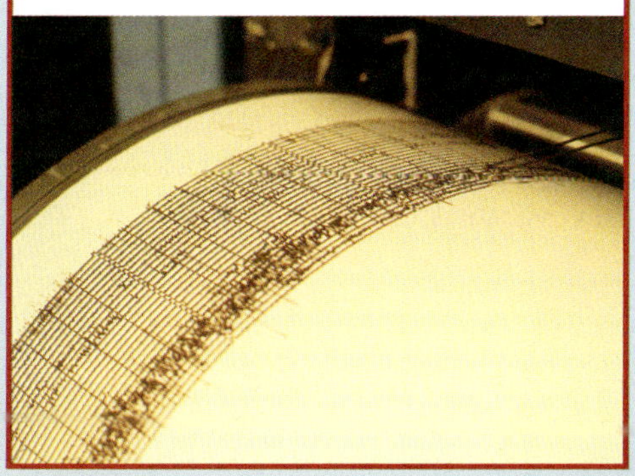

Evacuation Map

This map can be found in local Hawaiian phone books. It shows which areas must be evacuated when a tsunami is detected.

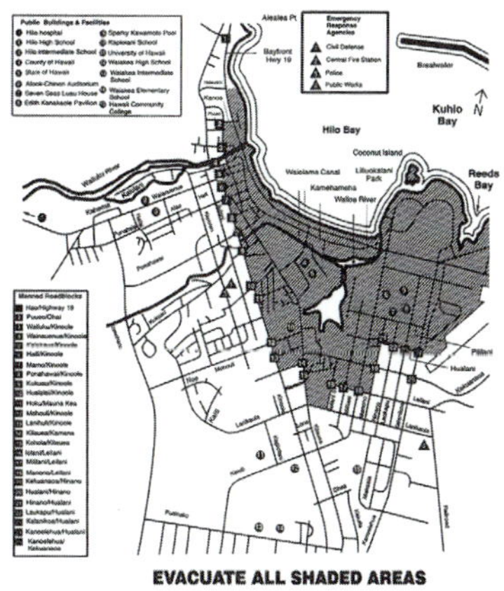

Warning Signal

Although tsunamis move very fast, warning sirens like this give people time to get to the safety of higher ground.

Chapter 2 Challenges

Chapter Summary

Lesson 2-1

- **Waves** are disturbances that transfer energy from place to place.
- Any substance through which waves can travel is called a **medium.**
- The particles of a medium do not move in the same direction as a wave but in small circles.

Lesson 2-2

- In a **transverse wave,** the particles of the medium move up and down at right angles to the direction of the wave motion. Its parts are the **crest** and the **trough.**
- In a **longitudinal wave,** the particles of the medium move back and forth in the same direction as the wave motion. Its parts are the **compression** and the **rarefaction.**

Lesson 2-3

- **Wavelength** is the distance from crest to crest or from trough to trough.
- **Frequency** is the number of complete waves passing a point in a given amount of time.
- **Amplitude** is the height of a wave.
- The **speed** of a wave is equal to the frequency multiplied by the wavelength.

Lesson 2-4

- A wave that strikes a barrier is called the **incident wave.**
- A wave that bounces off a barrier is called the **reflected wave.**
- The law of **reflection** states that the angle of incidence is equal to the angle of reflection.

Lesson 2-5

- The laws of **refraction** describe how waves are refracted, or bent, when they move from one medium to another.

Key Term Challenges

amplitude (p. 46)
compression (p. 44)
crest (p. 44)
electromagnetic wave (p. 42)
frequency (p. 46)
hertz (p. 46)
incident wave (p. 50)
longitudinal wave (p. 44)
mechanical wave (p. 42)
medium (p. 42)
normal (p. 50)
rarefaction (p. 44)
reflected wave (p. 50)
reflection (p. 50)
refraction (p. 52)
speed (p. 46)
transverse wave (p. 44)
trough (p. 44)
wavelength (p. 46)
wave (p. 42)

MATCHING **Write the Key Term from above that best matches each description.**

1. substance through which waves can travel
2. high point of a wave
3. height of a wave
4. unit used to measure frequency
5. wave that strikes a barrier
6. line at 90° to a barrier
7. disturbances in a medium

FILL IN **Write the Key Term from above that best completes each statement.**

8. In a __________, the particles of the medium move up and down at right angles to the direction of the wave motion.
9. In a __________, the particles of the medium move back and forth in the direction of the wave motion.
10. The two parts of a longitudinal wave are the __________ and the rarefaction.
11. The basic features of all waves are amplitude, __________, and frequency.
12. __________ occurs when a barrier does not absorb all of a wave's energy.

Content Challenges TEST PREP

MULTIPLE CHOICE Write the letter of the term or phrase that best completes each statement.

1. Waves are caused by
   a. potential energy.
   b. kinetic energy.
   c. heat energy.
   d. nuclear energy.

2. Mechanical waves cannot travel through
   a. air.
   b. water.
   c. metal.
   d. empty space.

3. When a longitudinal wave moves through a medium, the particles of the medium move
   a. in circles.
   b. up and down.
   c. back and forth.
   d. in the direction of the wave motion.

4. The two kinds of waves are transverse and
   a. circular.
   b. normal.
   c. longitudinal.
   d. compression.

5. The crest of a wave is the wave's
   a. low point.
   b. length.
   c. speed.
   d. high point.

6. In a transverse wave, the particles of the medium move
   a. at right angles to the direction of wave motion.
   b. back and forth to the direction of wave motion.
   c. at opposite angles to the wave motion.
   d. backward against the wave motion.

7. In a rarefaction, the particles are
   a. squeezed together.
   b. lined up.
   c. spread apart.
   d. not moving.

8. Wavelength can be measured in
   a. meters.
   b. hertz.
   c. angles.
   d. number of waves.

9. The wave's speed is equal to the frequency of the wave multiplied by the
   a. amplitude.
   b. wavelength.
   c. height.
   d. medium.

10. The angle between the normal and a barrier is equal to
    a. 45°.
    b. 90°.
    c. 180°.
    d. 360°.

TRUE/FALSE Write *true* if the statement is true. If the statement is false, change the underlined term to make the statement true.

11. The angle of incidence is equal to the angle of refraction.
12. When a wave is refracted, it is bent.
13. In a rarefaction, the particles are spread apart.
14. Mechanical waves do not need a medium to transfer energy.

Concept Challenges TEST PREP

WRITTEN RESPONSE **Answer each of the following questions in complete sentences.**

1. **HYPOTHESIZE:** You can make transverse waves in a rope tied to a doorknob. What happens to the waves when they reach the door?
2. **COMPARE:** How are the crests and troughs of a transverse wave like the compressions and rarefactions of a longitudinal wave?
3. **HYPOTHESIZE:** Suppose that light waves needed a medium through which to travel. How do you think the world would be different?

INTERPRETING A DIAGRAM **Use Figure 2-21 to answer the following questions.**

4. Which letter represents the wavelength of the wave?
5. Which letter represents the amplitude?
6. Which letter represents the frequency?
7. What is the relationship between speed, wavelength, and frequency?
8. If the frequency of a wave increases and the speed stays the same, what happens to the wavelength?
9. If the speed of a wave does not change but the wavelength increases, what happens to the frequency?

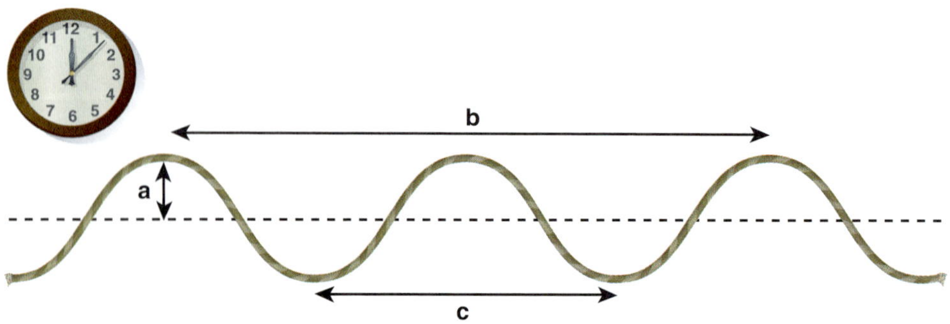

▲ **Figure 2-21**

Chapter 3 Sound

▲ **Figure 3-1** The instruments in this band make very recognizable sounds.

Sounds are all around you. When you hear a sound, most of the time you will recognize its source. For example, you would be able to tell the difference between the sound made by a bagpipe and one made by a bass drum. Although sounds are different, they all have something in common. All sounds are produced by objects that are moving back and forth rapidly.

▶ How does striking a drum produce a sound?

Contents

3-1 What is sound?

Objective
Identify sound as a form of energy caused by vibrations.

Key Terms
sound: form of energy that travels as waves
vibration: rapid back-and-forth movement

Sound and Energy There are sounds all around you. Some sounds are loud and others are very faint. Walking in a park, you may hear the sounds of birds singing and dogs barking. In a quiet room, you may hear the sound of an alarm clock ticking. Sound is a form of energy. **Sound** energy travels in the form of waves.

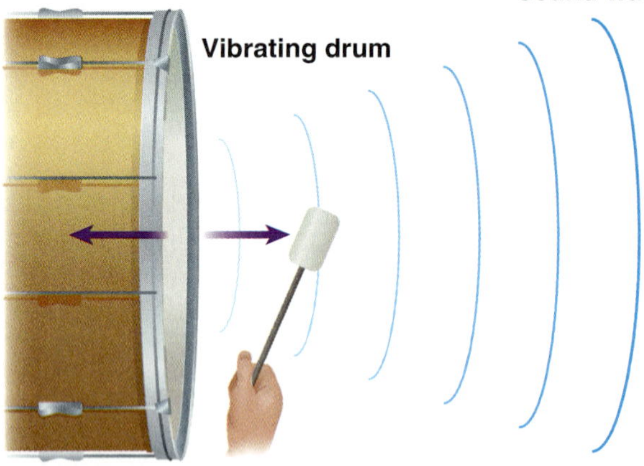

▲ **Figure 3-2** When a drum is struck, it vibrates and causes the air around it to vibrate.

1 DEFINE: What is sound?

Vibrations All sounds are caused by vibrations. A **vibration** is a rapid back-and-forth movement. Suppose you are listening to the sound from a stereo speaker. If you place your hand on the speaker, you will feel a vibration.

Objects that vibrate produce sound. Place your fingers on the front of your throat. When you speak, you can feel something vibrate. That something is your vocal cords (see Figure 3-3). The sound of your voice is produced when your vocal cords cause the air around them to vibrate.

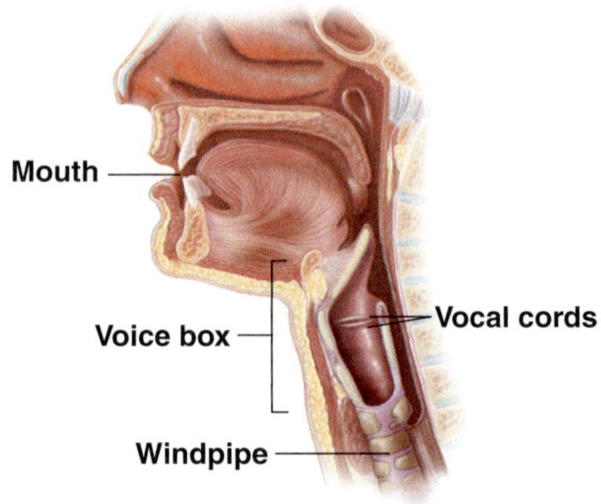

▲ **Figure 3-3** The sound of your voice is caused by the vibrations of your vocal cords.

2 DEFINE: What is a vibration?

Forming Sound For every sound that you hear, some object is vibrating. Strum a guitar. The guitar strings vibrate. When a string vibrates in one direction, it pushes on the air on that side of it. Air particles are squeezed together. The squeezed-together particles form a compression. When the string moves back, those air particles spread apart. The spread-out particles form a rarefaction. As the string continues to vibrate, compressions and rarefactions move away from the string. They form a sound wave.

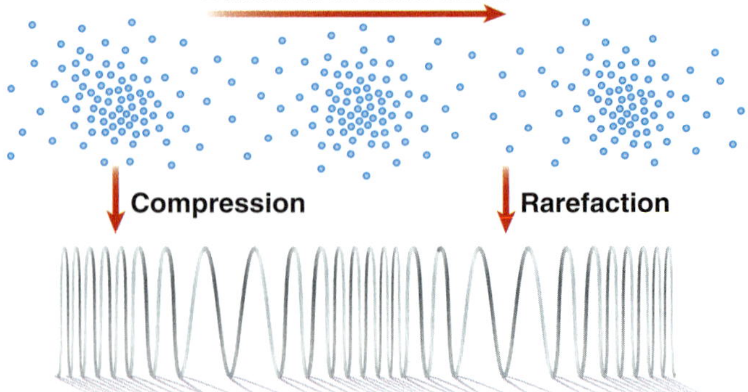

▲ **Figure 3-4** Sound waves are made up of a series of compressions and rarefactions.

3 DESCRIBE: What forms a sound wave?

Sources of Sounds You now know that when you hear a sound, something is vibrating to produce it. Different sounds can come from different sources. Figure 3-5 shows sources of a few of the sounds you might hear every day.

SOURCES OF SOUNDS	
Sound	**Vibration Source**
Violin	Strings
Drum beat	Drum skin and sticks
Human voice	Vocal cords
Knocking door	Door
Car revving	Engine
Pencil scratching	Lead and paper

▲ **Figure 3-5**

4 **DESCRIBE:** How many other sounds and sources can you add to the list in Figure 3-5?

CHECKING CONCEPTS

1. A sound is caused by a __________.
2. Sound is a form of __________ that travels as waves.
3. A __________ is a rapid back-and-forth movement.
4. All vibrating objects produce __________.
5. Sound waves are formed by compressions and __________.
6. If you place your fingers on your windpipe while you are speaking, you will feel a __________.
7. When you strum a guitar, the __________ vibrate.

THINKING CRITICALLY

8. **PREDICT:** What will happen if you place a glass of water on a stereo speaker?
9. **HYPOTHESIZE:** Is it possible to make a sound without causing a vibration? Why or why not?
10. **OBSERVE:** Sit by an open window. What sounds do you hear? What is causing each sound?

DESIGNING AN EXPERIMENT

Design an experiment to solve the following problem. Include a hypothesis, variables, a procedure, and a type of data to study.

PROBLEM: Can sound travel through solid objects?

Hands-On Activity

OBSERVING VIBRATIONS IN A TUNING FORK

You will need a tuning fork and a glass of water.

1. Place the glass of water on a flat surface. Let the glass sit for a few seconds until the water is still.
2. Gently strike the tuning fork against the edge of a table. Observe what happens.
3. Strike the edge of the table again with the tuning fork. Touch the ends of the tuning fork onto the surface of the water. Observe what happens to the water in the glass.

▲ **STEP 3** Put the ends of the vibrating tuning fork in the water.

Practicing Your Skills

4. **OBSERVE:** What happened to the tuning fork when you struck it against the table? Did you hear a sound?
5. **OBSERVE:** What happened when you put the tuning fork into the glass of water?

LAB ACTIVITY

Experimenting With Vibrations

BACKGROUND

Sound is created when an object vibrates. The vibrations are converted to waves that travel through the air. Different speeds of vibration produce different sounds. What is the relationship between the rate of vibration and the highness or lowness of the sound you hear?

PURPOSE

You will experiment with vibrations and sound, and create a simple musical instrument.

PROCEDURE

1. Copy the chart in Figure 3-6 on a separate sheet of paper.

2. Place a plastic ruler on a table so that the end of the ruler extends 5 cm past the edge of the table. Hold the ruler on the table with one finger at the 10-cm mark.

3. With your other hand pull up on the end of the ruler and release it. This action will cause the ruler to vibrate. Listen to the sound it produces. Record your observations in the chart.

4. Press down on the ruler at the 9-cm mark with a second finger. Again, pull up and release the end of the ruler and listen to the sound. Watch how fast the ruler vibrates.

5. Move the second finger to the 8-cm mark and again pull up and release the ruler. Repeat with your finger on the 7-cm and the 6-cm marks.

▲ **STEP 2** Hold a ruler near the edge of a table.

▲ **STEP 3** Make the ruler vibrate.

6. Try other positions with the ruler to make different sounds. Extend the ruler out further over the edge or bring it closer.

7. Use the data in your chart to write the "finger positions" as your "musical notes." Have a partner try to play your "song" from your written notes. See if you can play someone else's "song."

▲ **STEP 7** Play someone else's song.

Changing Sound

Trial	Finger Position in Centimeters	Speed of Vibration (very slow, medium, very fast, etc.)	Highness or Lowness of Sound (lowest, medium, highest, etc.)
1	10		
2	9		
3	8		
4	7		
5	6		
6			
7			
8			

▲ **Figure 3-6** Copy this chart and use it to record your observations.

CONCLUSIONS

1. **OBSERVE:** When the part of the ruler that vibrates is long, are the vibrations slow or fast?
2. **OBSERVE:** When the part of the ruler that vibrates is short, are the vibrations slow or fast?
3. **INFER:** If you hear a low sound, such as the sound produced by a foghorn, is it caused by slow vibrations or fast vibrations?
4. **INFER:** What kind of sound is produced by an object that vibrates fast?

3-2 How do sound waves travel?

INVESTIGATE

Observing Sound in a String Phone

HANDS-ON ACTIVITY

1. Make a small hole in the bottom of each of two paper cups.
2. Insert one end of a 10-m piece of string through the hole in each cup and tie a large knot at each end of the string.
3. Take one of the cups and give your partner the other. Move apart until the string is tight.
4. Have your partner speak into one cup while you hold the other cup to your ear. Take turns speaking into the cups.

THINK ABOUT IT: How does the sound travel from one cup to the other?

Objective

Describe how sound waves travel.

Key Terms

longitudinal (lahn-juh-TOOD-uhn-uhl) **wave:** wave in which the particles of the medium move back and forth in the direction of the wave motion

medium: material through which mechanical waves can travel

echo: reflected sound waves

Sound Waves Sound waves are longitudinal waves. In a **longitudinal wave,** the particles of the medium move back and forth in the direction of the wave motion. Figure 3-7 shows how a sound wave moves through the air. Compressions and rarefactions of air particles move in the same direction as the sound waves.

1 **DESCRIBE:** Why are sound waves longitudinal waves?

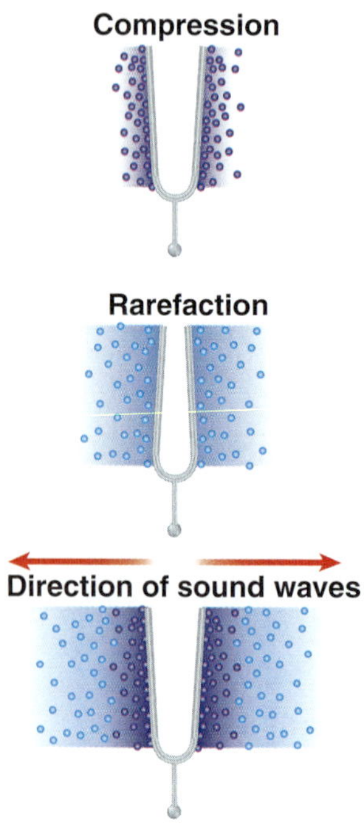

▲ **Figure 3-7** Parts of a sound wave.

Medium for Sound A **medium** is a material through which mechanical waves can travel. Air is a medium for sound waves. All sound waves need a medium through which they can travel. Sound waves can travel through gases, liquids, and solids. In each medium, the sound waves are made up of compressions and rarefactions of the particles that make up the medium.

Dolphins and whales use sound to communicate with each other. They can do this because sounds travel faster through water. Sounds also travel faster through solids. If you put your ear against a wall, you may hear sounds in the next room.

2 **EXPLAIN:** Why is water a good medium for communication between dolphins?

No Sound in a Vacuum On the Moon, astronauts cannot use sound to communicate. There is no air on the Moon. In a vacuum, there are no particles of a medium to compress. This means sound waves cannot be produced.

3 **INFER:** Would astronauts on the Moon be able to hear a rocket engine?

▲ **Figure 3-8** There is no sound on the Moon.

Reflected Waves Reflection is the bouncing back of a wave from a barrier that does not absorb its energy. If you clap your hands in a carpeted room, the sound will be absorbed by the carpeting. Suppose you clap your hands in an empty room with a wooden floor. You will hear the sound of the clap reflected from the floor and the walls. When a sound wave is reflected, the reflected sound wave is called an **echo.**

4 **IDENTIFY:** What is a reflected sound wave called?

CHECKING CONCEPTS

1. In a __________ wave, the particles of the medium move in the same direction as the wave motion.
2. A sound wave needs a __________ in which to travel.
3. Sound waves are made up of moving compressions and __________.
4. Whales can use sounds to communicate because sound waves can travel through __________.
5. Sound cannot be heard in a vacuum because there is no __________ for the sound waves.
6. Sound can travel through liquids such as water, gases such as air, and __________ such as walls.
7. Sound waves that are reflected from a surface are called __________.
8. When a sound wave is not reflected, the energy of the sound has been __________ by the surroundings.

THINKING CRITICALLY

9. **INFER:** If you clapped your hands on the Moon, would you hear the sound? Explain.

Web InfoSearch

Echolocation The biggest difference between bats and other animals is the way that they see. Most animals see by using their eyes. However, most bats "see" by using a system called echolocation.

SEARCH: Use the Internet to find out more about echolocation. Then, create a poster showing how echolocation in animals works. Start your search at www.conceptsandchallenges.com. Some key search words are **echolocation, bat sonar,** and **bat sight.**

Science and Technology

SONAR

Echoes can be used to measure distances under water. This method of using echoes is called sonar. The word *sonar* stands for **so**und **n**avigation **a**nd **r**anging. Sound waves are sent from a transmitter on a ship to the bottom of the ocean. The time it takes for the sound wave to reach the bottom and bounce back as an echo is measured. This time is then divided in half to find out how long the sound took to go one way. Suppose sound waves were sent out and returned in 4 seconds. The time for the sound waves to go one way is 2 seconds. The speed of sound in water is about 1,500 m/s. Therefore, the water is about 3,000 m deep. Sonar can be used to make maps of the ocean floor. It is also used to locate objects, such as boats and fish.

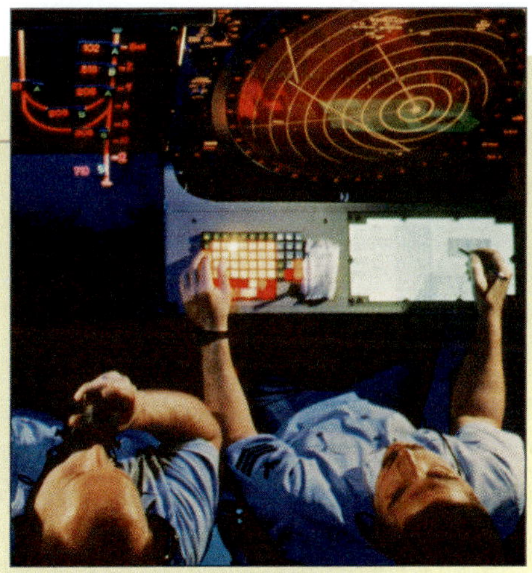

▲ **Figure 3-9** A sonar screen shows the locations of echoes.

Thinking Critically How do you think sonar can be used to find schools of fish?

3-3 What is the speed of sound?

Objective

Describe how the speed of sound changes in different materials.

Key Terms

supersonic: faster than the speed of sound for that medium

sonic boom: loud noise caused by breaking the sound barrier

Speed of Sound The speed of sound in air is about 340 m/s. This speed is much slower than the speed of light. In a thunderstorm, a flash of lightning is seen before the sound of thunder is heard. The light from the lightning flash travels much faster than the sound of the thunder produced by the lightning.

▲ **Figure 3-10** Lightning causes thunder.

You can use the speed of sound to estimate how far away a lightning flash is from your location. Start counting the seconds as soon as you see a flash of lightning. Stop counting when you hear the thunder. Sound travels about 1 km in 3 seconds. If it takes 3 seconds for the sound of the thunder to reach you, the lightning is about 1 km away.

1 STATE: What is the speed of sound in air?

Speed of Sound in Different Materials Sound travels at different speeds in different materials and at different temperatures. In water, sound travels at about 1,500 m/s. Figure 3-11 lists the approximate speeds of sound in some different materials.

SPEEDS OF SOUNDS AT 25°C	
Material	**Speed (m/sec)**
Air	346
Cork	500
Fresh water	1,498
Steel	5,200
Rubber	60
Glass	4,540
Wood (oak)	1,850

▲ **Figure 3-11**

2 ANALYZE: In which of the materials listed in Figure 3-11 does sound travel fastest?

Faster than Sound Many jet airplanes can travel faster than the speed of sound. Some airplanes can travel two or three times the speed of sound. Any speed faster than the speed of sound is said to be **supersonic.** When an airplane travels at supersonic speeds, it moves faster than the sound produced by its engines. When a plane traveling at supersonic speed passes overhead, it will be some distance away before you hear the sound of its engines. Figure 3-12 shows the first airplane, the Glamorous Glennis, to travel faster than the speed of sound. When something moves through air at speeds faster than the speed of sound, a loud noise is created. This noise is called a **sonic boom.**

▲ **Figure 3-12** Chuck Yeager piloted the first plane to travel at supersonic speeds.

In 1997, Andy Green broke the sound barrier on land with his car named Thrust. He did so exactly fifty years after the sound barrier was broken by Captain Chuck Yeager. Figure 3-13 shows the supersonic car that Green used to accomplish this feat.

▲ **Figure 3-13** Andy Green's supersonic car

3 **DEFINE:** What is supersonic speed?

CHECKING CONCEPTS

1. The speed of sound in air is about __________.
2. You can see a __________ airplane before you hear it because it is moving faster than sound.
3. Light travels __________ than sound.
4. A flash of lightning can be seen __________ the sound of thunder it produces can be heard.

THINKING CRITICALLY

Use the information in Figure 3-11 to answer the following questions.

5. **COMPARE:** Would the sound of an approaching train travel faster through the metal railroad tracks or through the air? Explain.
6. **HYPOTHESIZE:** **a.** Why do you think rubber is sometimes used for soundproofing a room? **b.** What other materials in the table would be good for soundproofing? Explain.

BUILDING MATH SKILLS

Calculating *Find out how long it would take sound to travel through 2 m of the following materials.*

a. air
b. water
c. steel
d. rubber
e. glass
f. wood
g. cork

Science and Technology

SUPERSONIC AIRPLANES

The speed of sound in air is called Mach 1. Before October 14, 1947, no airplane had ever reached the speed of Mach 1. On that day, Chuck Yeager became the first person to fly faster than the speed of sound. He had broken the sound barrier.

Today, supersonic planes can fly at speeds of Mach 2 or even Mach 3. For example, the supersonic *Concorde* crosses the Atlantic Ocean in much less time than other passenger airplanes.

▲ **Figure 3-14** The *Concorde* is an airliner that can travel at supersonic speeds.

One disadvantage of planes that travel at supersonic speeds is that they produce loud noises called sonic booms. A sonic boom is caused when a plane breaks the sound barrier. Sonic booms can be so loud that they break windows and knock pictures off walls. To avoid the problem of sonic booms, supersonic planes are not allowed to fly over heavily populated areas.

Thinking Critically The *Concorde* only travels at supersonic speeds over the ocean. Why do you think this is so?

3-4 What is intensity?

Objective

Explain the relationship between the intensity and the loudness of a sound.

Key Terms

intensity: amount of energy in a sound wave

decibel: unit used to measure the intensity or loudness of a sound

Intensity Sound energy travels in the form of waves. Some sounds are loud. Others are soft. Figures 3-15 and 3-16 show wave patterns for a soft sound and a loud sound.

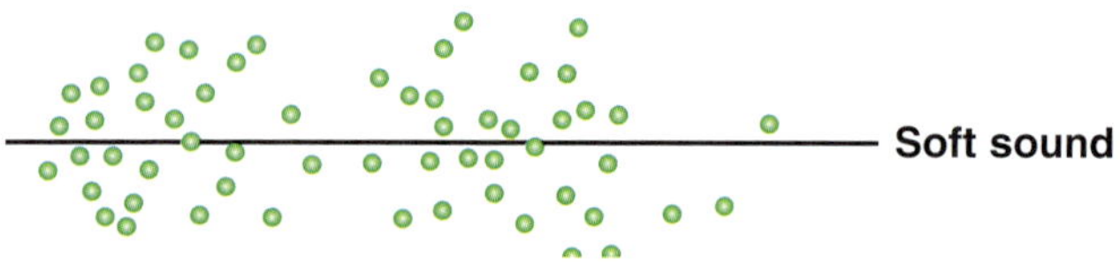

▲ **Figure 3-15** Wave pattern of soft sound

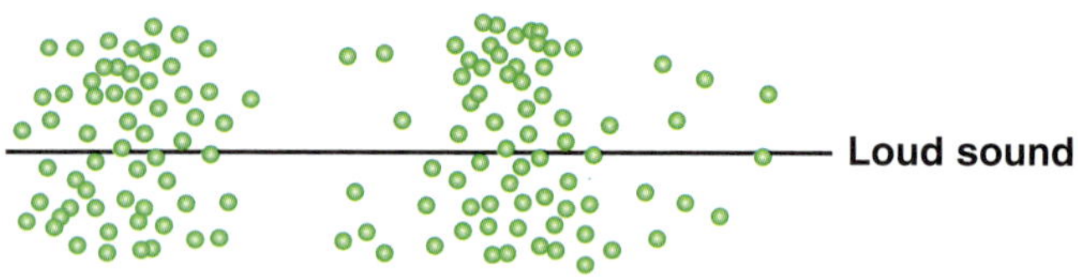

▲ **Figure 3-16** Wave pattern of loud sound

The amount of energy a sound has is called the **intensity** of the sound. As Figures 3-15 and 3-16 show, a wave carrying a loud sound has a greater amplitude than that of a wave carrying a soft sound. So, molecules in a wave carrying a loud sound travel farther than do the molecules in a wave carrying a soft sound. The molecules in the loud wave have more energy.

▲ **Figure 3-17** Which instrument will produce sounds of greater intensity?

1 **DEFINE:** What is intensity?

Measuring Intensity Intensity is measured in units called **decibels** (dB). The sound of people talking has an intensity of about 65 decibels. The softest sound a human can hear is about 20 decibels. The sound of a train going by may have an intensity of 95 decibels. Sounds louder than 120 decibels can be dangerous. They can damage your ears. Figure 3-18 shows the intensity of different sounds measured in decibels.

INTENSITY OF SOUND

Sound	Intensity (dB)
Cat purring	10–20
Soft music	30
Average home	40–50
Conversation	60–70
Heavy traffic	70–80
Loud music	80–100
Thunder	110
Rock concert	115–120
Jet engine	170
Rocket engine	200

▲ **Figure 3-18**

2 **ANALYZE:** What is the intensity of the sound at a rock concert?

CHECKING CONCEPTS

1. When you raise the volume on a radio, you are increasing the __________ of the sound waves.
2. The __________ in a sound wave determines the amplitude of the wave.
3. The __________ of a sound wave is the amount of energy it has.
4. The unit used to measure intensity is the __________.
5. The more intensity a sound wave has the __________ the sound.

THINKING CRITICALLY

6. **INFER:** Which is more dangerous to your ears, heavy traffic noise or a rock concert? Explain.
7. **HYPOTHESIZE:** Two people hear a loud sound. One person is 10 m from the source of the sound. The other person is 50 m from the source. The sound does not seem as loud to the person 50 m from the source as it does to the person standing closer. What does this tell you about the intensity of a sound wave?

HEALTH AND SAFETY TIP

Your ears are very sensitive to sound. Loud sounds can damage your hearing. An intensity of 120 db is called the threshold of pain. This means that a sound louder than 120 dB is so loud that it will hurt your ears. You should keep this in mind when listening to live music or when using earphones to listen to a recording. Use library references to find out how to protect yourself from dangerously loud sounds. Write a short report on your findings.

Integrating Environmental Science

TOPIC: noise pollution

TOO MUCH SOUND

The world is full of sounds. There are many sources of sound. Some of the loudest sounds come from machines. These loud machine sounds may cause noise pollution. Any loud, unwanted sound is called noise. In cities, the sounds of street traffic, of machines at construction sites, and of airplanes all combine to cause noise pollution.

Noise pollution is a nuisance but it can also be a hazard. Too much noise can damage the ears by destroying the delicate hairs and tiny muscles in the ears. Noise can also cause stress. People who live or work around too much noise may develop high blood pressure or a stress-related illness.

There are ways to decrease noise pollution. People are encouraged to use mass transit to decrease the amount of traffic on city streets. Laws have been passed to keep people from playing radios too loudly in public places. You are encouraged to use ear protection when exposed to loud noises over an extended period of time.

▲ **Figure 3-19** Construction work contributes to noise pollution.

Thinking Critically What can you do to help prevent noise pollution?

3-5 What are frequency and pitch?

Objective

Explain how frequency and pitch are related.

Key Terms

pitch: how high or low a sound is

ultrasonic (uhl-truh-SAHN-ihk)**:** sound above 20,000 Hz frequency

infrasonic: sound below 20 Hz frequency

Frequency of a Sound Different sounds have different frequencies. The frequency of a wave is the number of complete waves that pass a point each second. Each pair of compressions and rarefactions is a complete sound wave. The frequency of a sound wave is the number of compressions and rarefactions produced per second. Frequency is measured in hertz (Hz). One hertz is equal to one wave per second.

1 IDENTIFY: What is frequency?

Frequency and Pitch Frequency and pitch are related. A sound's **pitch** tells how high or low a sound is. A high-frequency sound has a high pitch. A low-frequency sound has a low pitch. A toy whistle has a frequency of about 1,000 Hz. This is a high-pitched sound. Thunder, a low-pitched sound, has a frequency of about 50 Hz.

▲ **Figure 3-20** These instruments can make both high- and low-pitched sounds.

2 DESCRIBE: How are the frequency and the pitch of a sound related?

Changing Pitch A guitar can make sounds of many different pitches. If you look at the strings of the guitar in Figure 3-21, you will see that some are thick and some are thin. The thick strings make low-pitched sounds. The thin strings make high-pitched sounds. A guitar has tuning pegs that tighten or loosen the strings. Changing the tightness of the strings also changes the pitch. Placing your fingers on the frets across the neck of the guitar changes the length of the string. This action changes the pitch of the sound produced.

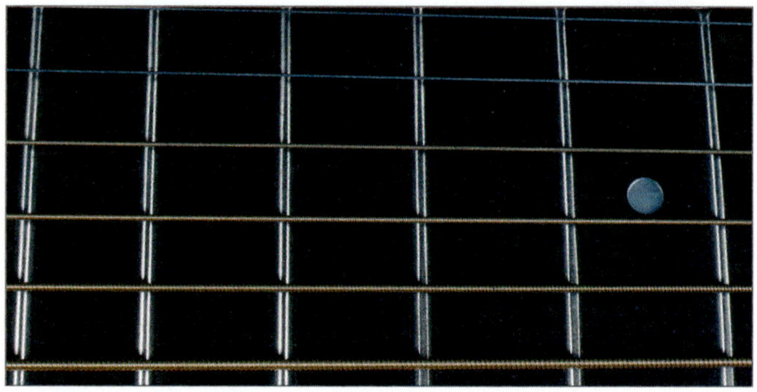

▲ **Figure 3-21** The different strings on a guitar make sounds of different pitch.

3 HYPOTHESIZE: How does changing the tightness of a string affect the sound it produces?

Range of Hearing Humans can hear sounds with frequencies between 20 Hz and 20,000 Hz. This is called the range of human hearing. Sounds that have frequencies higher than 20,000 Hz are called **ultrasonic.** The prefix *ultra-* means "above." Ultrasonic sounds have too high a pitch to be heard by humans. The range of human hearing decreases as a person gets older. Children can hear higher-frequency sounds better than adults can.

Some animals can hear ultrasonic sounds. Dogs can hear sounds with frequencies up to about 50,000 Hz. Often, dogs are trained using ultrasonic whistles. The dogs hear it but people do not.

4 DEFINE: What is an ultrasonic sound?

Infrasonic Sound Sound that is too low for humans to hear is called **infrasonic.** The prefix *infra-* means "below." Infrasonic is any sound below 20 Hz.

Elephants communicate distress with infrasonic sound at about 6 Hz. They stamp on the ground with their feet. The sound produces waves that travel many miles to other elephants.

◀ **Figure 3-22** Elephants communicate with infrasonic sounds.

5 **DEFINE:** What is an infrasonic sound?

CHECKING CONCEPTS

1. The unit used to measure frequency is the __________.
2. Sounds that humans can hear are between 20 and __________ Hz.
3. Even if two sounds have the same intensity, they can have different __________.
4. One hertz is equal to one __________ per second.

THINKING CRITICALLY

5. **COMPARE:** Which sound has a lower pitch, a sound with a frequency of 20 Hz or one with a frequency of 20,000 Hz? How do you know?
6. **HYPOTHESIZE:** The sound of a siren coming toward you has a higher pitch than the sound of a siren that is standing still. What does this tell you about the frequency of a sound that is moving toward you?

Web InfoSearch

Ultrasound A technique called ultrasound can be used to see inside solid objects. Ultrasound uses very high-pitched sounds. Most of the sound waves pass through the object, but some are reflected. The reflected waves produce an image on a screen. Ultrasound images are similar to X-rays.

SEARCH: Use the Internet to find out how ultrasound is used in the fields of medicine, engineering, and construction. Start your search at www.conceptsandchallenges.com. Some key search words are **ultrasound, medicine, engineering,** and **construction.**

Hands-On Activity

OBSERVING CHANGES IN PITCH

You will need 5 test tubes, a test-tube rack, a metric ruler, and a container of water.

1. Place the test tubes in the test-tube rack.
2. Pour different amounts of water into each test tube. Use the metric ruler to measure the height of the column of air above the water in each test tube. The heights should be as follows: 2 cm, 4 cm, 6 cm, 8 cm, and 10 cm.
3. Blow across the top of each test tube. Observe the pitch of the sound you hear from each tube.

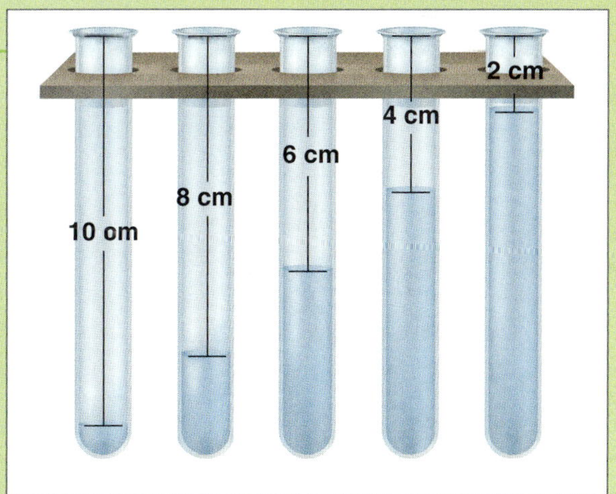

▲ **STEP 2** Set up the test tubes so each has a different amount of air above the water.

Practicing Your Skills

4. **INFER:** How does the height of the column of air affect the pitch of the sound?
5. **OBSERVE:** Which test tube produced the highest pitch?
6. **OBSERVE:** Which test tube produced the lowest pitch?

Integrating Life Science

THE Big IDEA

How do animals use sound?

Listen to the world around you. Vibrations in the air make your eardrum move. Tiny bones in your middle ear amplify the vibrations. Then, your inner ear changes the vibrations into messages for your brain. Human ears work in the range of about 20 to 20,000 vibrations per second or hertz.

Some animals make sounds that are too high- or too low-pitched for our ears to detect. Sounds above the range of human hearing are called ultrasonic and sounds below the range of human hearing are called infrasonic.

Bats make ultrasonic chirps with a wide range of frequencies. They use the echoes of their chirps to locate food with sonar. Dolphins and other sea mammals also use ultrasound for communication and sonar.

Humpback whales are amazing vocalists. They sing mostly in the autumn before migrating to tropical feeding grounds. Each group's song changes from season to season. Humpback songs can last for 30 minutes and range from 20 to 9,000 Hz.

Elephants communicate with infrasonic sounds that travel many kilometers. When some family groups of elephants come together, they flap their ears and exchange greetings in the range of 5 to 50 Hz, a very low-pitched way of saying hello.

Look at the illustrations that appear on these two pages. Then, follow the directions in the Science Log to find out more about "the big idea." ✦

Elephants
When elephants greet each other they stand side by side and exchange rumbling infrasonic hellos that vary from 5 to 50 Hz.

Whales
Humpback whales sing complex eerie sounding songs that range from 20 to 9,000 Hz. Long ago sailors believed the songs were the cries of sailors' ghosts.

Bats

Bats are so good at using ultrasound to locate that they can catch moths and mosquitoes even when they are flying at full speed. Bat chirps have a frequency from 10,000 to over 100,000 Hz.

WRITING ACTIVITY

Science Log

On the Web find the sound of any animal. Listen carefully to it. Write a hypothesis predicting what the sound means for the animal. How could you test your hypothesis? Write down a procedure for an observation or an experiment that would help you test your hypothesis on animal sounds. For additional information for your experiment, visit www.conceptsandchallenges.com.

Wolves

Some researchers believe that wolves can hear sounds all the way up to 80,000 Hz. Wolves have many calls. The wolf's "lonesome" call ranges from 200 to 2,200 Hz.

Dolphins

Bottlenose dolphins make sounds for echolocation and communication. Each task has a different range of sound, varying from about 250 Hz to 150,000 Hz. The frequency during echolocation is about 100,000 Hz. Many people wonder if they have languages like humans.

3-6 What is sound quality?

INVESTIGATE

Listening to a String Instrument

HANDS-ON ACTIVITY

STEP 3

1. Pluck the lowest-pitched string of a violin or guitar.
2. Lightly touch the string in the center to change its vibrations. LIsten for the higher note.
3. Pluck the same string again. Touch it one-third of the way along its length to get the next highest note.

THINK ABOUT IT: How does the pitch and loudness of these sounds affect the quality of the sound heard?

Objective

Explain how overtones affect sound quality.

Key Terms

timbre (TAM-buhr)**:** sound quality

fundamental tone: lowest pitched sound produced when an object vibrates

overtones: sounds that have a higher pitch than the fundamental tone

music: a pleasing combination of sounds

noise: unpleasant combination of sounds with irregular patterns of vibration

Sound Quality Suppose a trumpet player and a clarinet player play the same note. The sounds they make have the same frequency and the same intensity. Can you hear a difference between the two sounds? The sounds are different even though the frequency and the intensity are the same. The sounds differ because of their **timbre,** or sound quality. The sounds of a trumpet and a clarinet have different timbres.

1 DEFINE: What is timbre?

Fundamental Tone When an object vibrates, it produces a sound with a certain frequency. Figure 3-23 shows a vibrating string. Notice that the whole string is vibrating. When the whole string vibrates, it produces the lowest possible frequency and pitch. The lowest-pitched sound produced is called the **fundamental tone.**

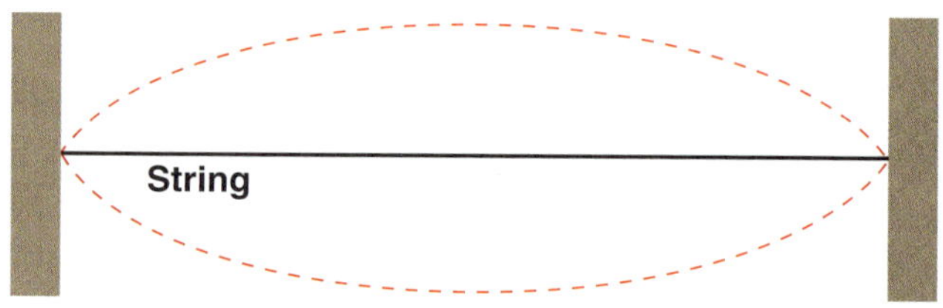

▲ **Figure 3-23** Fundamental tone

2 DEFINE: What is a fundamental tone?

Overtones Parts of the string shown in Figure 3-24 can vibrate faster than the whole string. When parts of the string are vibrating, the string produces sounds with a higher pitch than the fundamental tone. The high-pitched sounds produced when different parts of a string vibrate are called **overtones.**

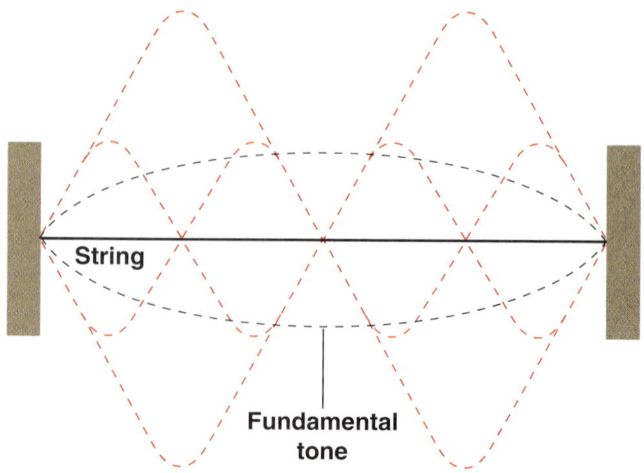

▲ **Figure 3-24** The dashed red lines are overtones.

The sounds produced by a musical instrument or a human voice also have overtones. The quality of a sound is a result of these overtones. Without overtones, all vibrations of the same frequency would have the same quality.

3 **DEFINE:** What are overtones?

Music Different musical instruments have different sound qualities. How do the different instruments in an orchestra combine to make music? **Music** is a pleasing combination of sounds. Some elements of music are rhythm, melody, and harmony.

- **Rhythm** The basic beat in music is known as the rhythm. When you listen to music, the rhythm you hear might come from drums or a bass guitar. If you clap your hands to the beat of the music, you are adding rhythm.
- **Melody** The combination of musical notes is called the tune or the melody. If you hum or sing a song, you are repeating the melody.
- **Harmony** In an orchestra or chorus, different instruments or voices perform together. Many different notes are heard at the same time. Combining the different notes is called harmony. Good harmony makes music sound better.

4 **IDENTIFY:** What are three elements of music?

Musical Instruments Musical instruments can be divided into three main groups.

- **Percussion Instruments** These instruments include drums, pianos, and any instrument that is played by being tapped or hit.
- **String Instruments** These instruments make music from the sound of vibrating strings. Examples include guitars, violins, and cellos.
- **Wind Instruments** These instruments make music when air is blown through them. Examples include trumpets, clarinets, and flutes.

5 **NAME:** What are the three main groups of musical instruments?

Noise **Noise** is an unpleasant combination of sounds with irregular patterns of vibration. Unlike music, noise has no agreeable quality.

6 **DEFINE:** What is noise?

CHECKING CONCEPTS

1. The quality of a sound is its ___________.
2. When a whole string vibrates, it produces a ___________.
3. All sounds are a combination of fundamental tones and ___________.
4. Two sounds can have the same ___________ and intensity but still sound different.
5. Higher-pitched sounds produced when parts of a string vibrate are called ___________.
6. Unpleasant sounds are ___________.
7. Rhythm, melody, and harmony combine to make ___________.

THINKING CRITICALLY

8. **INFER:** What would human voices sound like without overtones?
9. **INFER:** A violin and a piano are playing the same musical note with the same frequency and the same intensity. Why do they sound different?
10. **CLASSIFY:** Identify each of the following instruments as percussion, string, or wind instruments.
    - **a.** tuba
    - **b.** kettle drum
    - **c.** harp
    - **d.** trumpet

Web InfoSearch

The Musical Scale Our musical scale is made up of eight notes. Each musical note has its own pitch. Each note is represented by a letter of the alphabet. A set of eight notes is called an octave. The last note in an octave has twice the frequency of the first note. The human ear can hear a total of about 10 octaves.

SEARCH: Use the Internet to find out what a musical scale looks like. Draw a scale with all the notes placed properly. Start your search at www.conceptsandchallenges.com. Some key search words are **music** and **musical scale.**

3-7 What is the Doppler effect?

Objective

Explain what is meant by the Doppler effect.

Key Term

Doppler effect: apparent change in the frequency of waves

Changing Frequency The frequency of a wave sometimes changes. Remember that frequency is the number of complete waves that pass a point in a given time. Frequency seems to change when a wave source moves toward you or away from you. Imagine you are sitting on a dock counting the number of waves hitting the dock. As a motorboat passes you, many waves hit the dock. This means that the frequency of the waves is high. As the boat continues on, fewer waves hit the dock. The frequency returns to what it was before the boat passed.

▲ **Figure 3-25** The frequency of waves changes as a boat passes.

1 COMPARE: Will you be able to count more waves when a boat is approaching or heading away from a dock?

Doppler Effect An apparent change in the frequency of waves is called the **Doppler effect.** The Doppler effect occurs when there is relative motion between the source of the waves and an observer. The frequency of waves appears to change when the observer is moving toward or away from the source of the waves. The frequency also seems to change when the source of the waves is moving and the observer is standing still. For the Doppler effect to take place, either the source or the observer must be moving.

2 DEFINE: What is the Doppler effect?

Doppler Effect and Sound You are probably most familiar with the Doppler effect in sound waves. The frequency of sound waves seems to change as the source of the waves moves toward or away from you. Suppose you are waiting for a train to pass a crossing. You can hear the train whistle as the train approaches the crossing. The waves are pushed closer together by the motion of the train. The pitch of the sound seems to get higher as more and more waves per second reach your ears. As the train passes you, the sound waves spread out. Fewer waves reach your ear every second. The pitch of one sound seems to get lower.

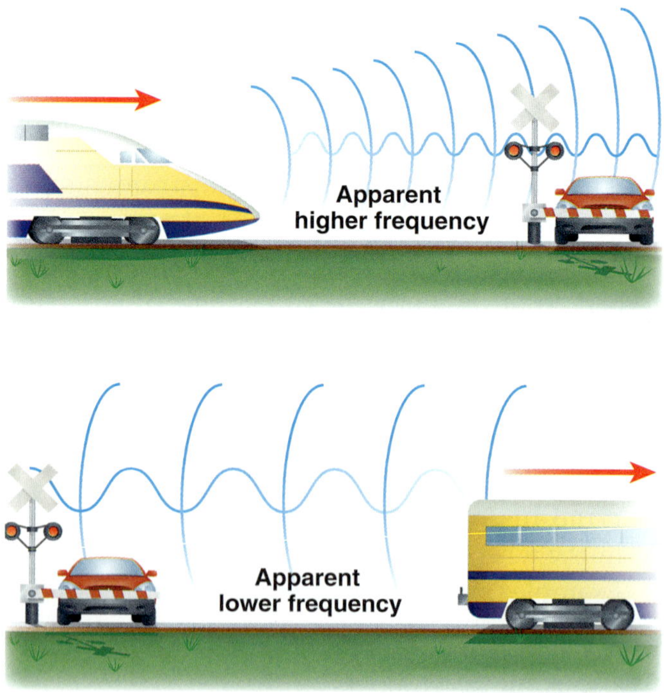

▲ **Figure 3-26** The Doppler effect causes the sound of a train whistle to change as it passes.

3 COMPARE: Does the frequency of sound waves from a train whistle become higher or lower as the train comes toward you?

CHECKING CONCEPTS

1. The apparent change in the frequency of waves is called the __________.
2. Waves hit a dock more often when a boat is moving __________ the dock.
3. The Doppler effect is caused by __________.
4. The frequency of waves appears to change when either the source of the waves or the __________ is moving.
5. The frequency of sound waves is __________ when the source of the waves is moving away from you.
6. As a boat moves toward you, the frequency of the water waves appears __________ than if the boat was moving away from you.

THINKING CRITICALLY

7. **EXPLAIN:** Describe how the sound of a car horn changes as the car approaches you and then passes you.

Use Figure 3-27 to answer question 8.

8. **PREDICT:** Will the pitch of the sound of the fire engine's siren seem higher to a person standing at point A or point B? Explain.

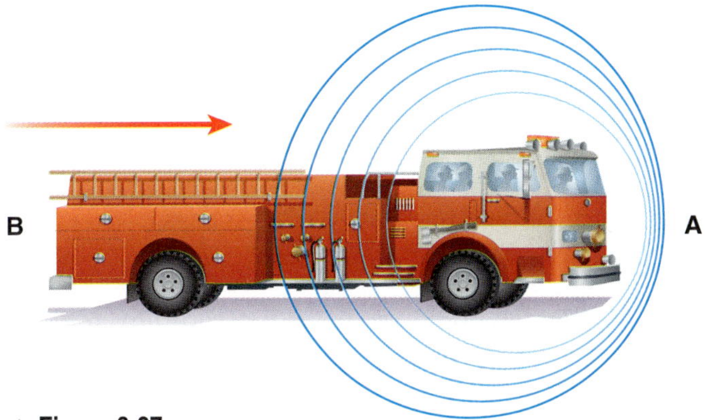

▲ **Figure 3-27**

BUILDING SCIENCE SKILLS

Researching Visit your local police station. Many police forces use Doppler radar to measure the speed of passing cars. Find out how Doppler radar works. How do radar detectors in cars work? How is Doppler radar used to track the path of fast-moving storms? What are other uses of Doppler radar? Describe your findings to the class.

Integrating Earth Science

TOPICS: stars, universe

RED SHIFT

Astronomers study the wavelength of light from the stars to find out about the universe. The wavelength of light determines the color of the light. Red light has a long wavelength. Blue light has a short wavelength.

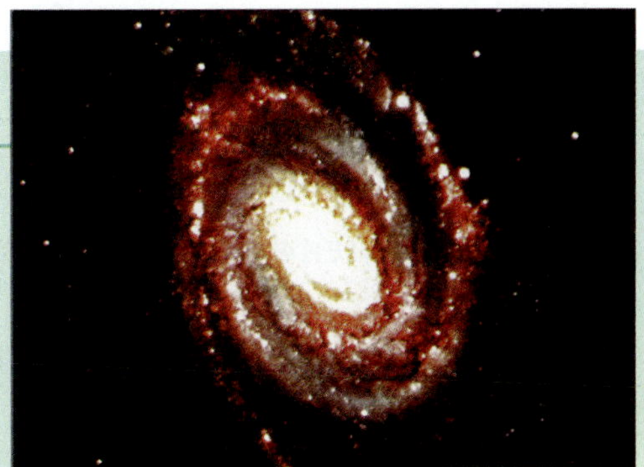

▲ **Figure 3-28** Galaxies are moving apart.

The Doppler effect causes an apparent change in the frequency of a wave. It also causes an apparent change in wavelength. If a light source is moving away from an observer, the wavelength of the light appears to change. The light appears redder than it would if the source were not moving. This change in wavelength is called the red shift.

Astronomers have studied the red shifts of many different galaxies. They found that all of the galaxies are moving away from Earth. Each galaxy is also moving away from every other galaxy. This suggests that the universe is expanding, or getting bigger.

Thinking Critically What would astronomers see if the universe were shrinking?

3-8 How do you hear?

Objective
Trace the path of sound waves through the ear to the brain.

Key Terms
hearing: one of the five human senses

ear: sense organ that detects sound

eardrum: thin sheet of tissue that vibrates when sound waves strike it

cochlea (KAHK-lee-uh)**:** organ that changes sound vibrations into nerve signals

Hearing Sounds One of the five human senses is **hearing.** What happens when you hear a sound? For a sound to be heard, three things are needed: a source of the sound, a medium to transmit the sound, and a sense organ to detect the sound.

In humans, the sense organ that detects sound is the **ear.** Look at the diagram of the human ear in Figure 3-29. There are three main parts to the ear: the outer ear, the middle ear, and the inner ear.

1 NAME: What are the three main parts of the human ear?

How the Ear Hears A sound wave first enters the outer ear. The outer ear funnels the sound wave into the ear. The sound wave moves through the ear canal to the eardrum. The **eardrum** is a thin sheet of tissue that vibrates when sound waves strike it. The vibrating air particles in the sound wave make the eardrum vibrate.

The vibrations from the eardrum are transferred to the middle ear. There are three small bones in the middle ear. They are the hammer, the anvil, and the stirrup. The vibrations are transferred from the hammer, to the anvil, and then to the stirrup.

The vibrations are then transmitted to the inner ear. In the inner ear, the vibrations are transferred to the cochlea. The **cochlea** is the organ that changes sound vibrations into nerve impulses. The cochlea is filled with liquid and is attached to nerve fibers. The nerve fibers join to form one nerve. The nerve transmits the impulses to the brain. In the brain, the nerve impulses are interpreted as sound.

2 NAME: Which parts of the ear vibrate when a sound is heard?

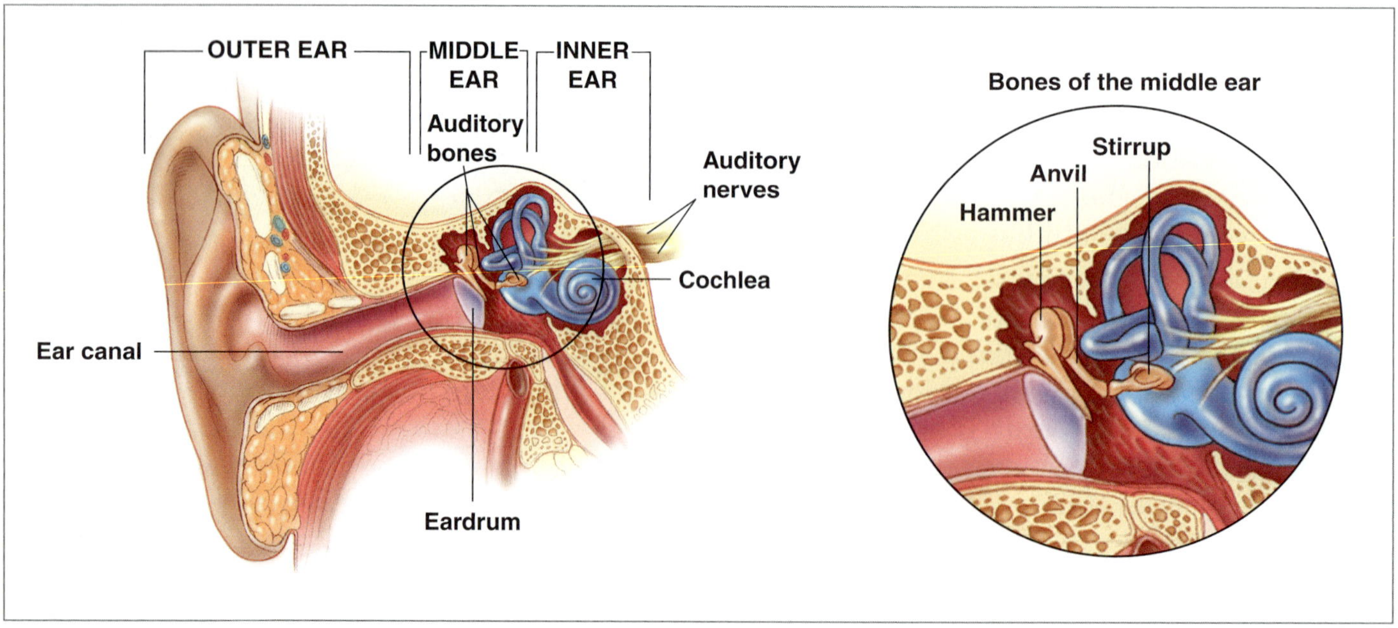

▲ **Figure 3-29** Parts of the human ear

Dizziness The inner ear has another important role that has nothing to do with hearing. The inner ear maintains balance. After you ride a roller coaster, you may feel dizzy. Dizziness is a temporary loss of balance. It is often caused by an ear infection. The infection irritates the tiny hair cells that project into fluid-filled canals in the inner ear's balance center. When irritated, these hair cells send mixed-up messages to the brain, tricking the brain into thinking you are spinning.

3 **INFER:** Why is it important to maintain balance?

CHECKING CONCEPTS

1. The __________ is the organ that detects sound.
2. Of the five senses, the one that is involved with sound is __________.
3. Sound waves travel from the outer ear, to the __________, and then to the inner ear.
4. The part of the ear that vibrates first is the __________.
5. The __________ converts the sound vibrations into nerve impulses.

THINKING CRITICALLY

6. **SEQUENCE:** Arrange the following steps in the correct order.
   a. The vibrations are transferred to the hammer, anvil, and stirrup. **b.** The vibrations of air particles in the sound wave cause the eardrum to vibrate. **c.** The vibrations are transferred to the cochlea. **d.** The brain interprets the electrical impulses as sounds. **e.** A sound wave is transmitted by the source. **f.** The vibrations are converted to electrical impulses along a nerve that goes to the brain. **g.** The sound waves travel through the medium. **h.** The sound waves enter the outer ear and go into the ear canal.

INTERPRETING VISUALS

Look at Figure 3-29 to help you answer the following questions.

7. Where is the liquid that helps you maintain balance located?
8. Which parts of the ear have a liquid medium for sound to travel?
9. Which parts of the ear have a solid medium?
10. Which parts of the ear have a gaseous medium?

Real-Life Science

THE CHANGING TELEPHONE

In 1876, Alexander Graham Bell received a patent for the first working telephone. Two years later, telephone service began in the United States. Since that time, the telephone has undergone many changes and improvements.

In 1977, communications companies began testing mobile, or cellular, telephones. By 1987, more than one million people had subscribed to cellular phone systems. Today, personal computers can be used to connect to regular telephones and can be used to make video phone calls.

▲ **Figure 3-30** Alexander Graham Bell testing his phone

The next dramatic change in the telephone will likely be disposable cellular telephones. Prototypes of these telephones have already been produced. They are as thin as a credit card and can be recycled.

THINKING CRITICALLY: What are some advantages of disposable phones?

Chapter 3 Challenges

Chapter Summary

Lesson 3-1

- **Sound** is a form of energy that travels as waves. All sounds are caused by **vibrations.**

Lesson 3-2

- Sound waves are **longitudinal waves.** All sound waves need a **medium** through which to travel. A reflected sound wave is called an **echo.**

Lesson 3-3

- The speed of sound is much slower than the speed of light. Sound travels at different speeds in different materials. Jet airplanes can reach supersonic speeds.

Lesson 3-4

- Loud sounds have more energy than soft sounds do. The **intensity** of a sound wave determines the loudness of the sound. Intensity is measured in **decibels** (dB).

Lesson 3-5

- Different sounds have different frequencies. A high-frequency sound has a high pitch. A low-frequency sound has a low pitch.
- Some animals can hear **ultrasonic** or **infrasonic** sounds.

Lesson 3-6

- The sounds of a trumpet and a clarinet differ because of their sound quality, or **timbre.**
- When a string vibrates as a whole, it produces a **fundamental tone.**
- When different parts of a string vibrate, they produce **overtones.**
- **Music** is a combination of rhythm, melody, and harmony. **Noise** is an unpleasant combination of sounds with no regular pattern of vibrations.

Lesson 3-7

- The **Doppler effect** is caused by motion of the observer or of the source of the waves.

Lesson 3-8

- The **ear** is the human sense organ that detects sound.

Key Term Challenges

cochlea (p. 78)
decibel (p. 68)
Doppler effect (p. 76)
ear (p. 78)
eardrum (p. 78)
echo (p. 64)
fundamental tone (p. 74)
hearing (p. 78)
infrasonic (p. 70)
intensity (p. 68)
longitudinal wave (p. 64)
medium (p. 64)
music (p. 74)
noise (p. 74)
overtone (p. 74)
pitch (p. 70)
sonic boom (p. 66)
sound (p. 60)
supersonic (p. 66)
timbre (p. 74)
ultrasonic (p. 70)
vibration (p. 60)

MATCHING **Write the Key Term from above that best matches each description.**

1. how high or low a sound is
2. speed faster than sound
3. rapid back-and-forth movement
4. frequency above the fundamental tone
5. sound quality
6. sense organ that detects sound
7. includes rhythm, melody, and harmony
8. one of the human senses
9. form of energy that travels as waves
10. unit of intensity

FILL IN **Write the Key Term from above that best completes each statement.**

11. A sound wave needs a ____________ in which to travel.
12. A reflected sound wave is an ____________.
13. The amount of energy in a sound wave is the ____________ of the wave.
14. Sound with a frequency greater than 20,000 Hz is ____________.
15. The ____________ is the lowest-pitched sound produced when a whole string vibrates.

Content Challenges TEST PREP

MULTIPLE CHOICE Write the letter of the term or phrase that best completes each statement.

1. To find their way in the dark, bats use
   a. overtones.
   b. ultrasound.
   c. echolocation.
   d. timbre.
2. A high-pitched sound has a
   a. low frequency.
   b. fundamental tone.
   c. high frequency.
   d. low intensity.
3. Two sounds with the same frequency and intensity will sound different because of
   a. timbre.
   b. pitch.
   c. vibrations.
   d. wavelength.
4. The speed of sound in air is
   a. 125 m/s.
   b. 340 m/s.
   c. 430 m/s.
   d. 540 m/s.
5. The unit of frequency is the
   a. hertz.
   b. decibel.
   c. m/s.
   d. pitch.
6. Sound travels fastest in
   a. liquids.
   b. solids.
   c. gases.
   d. a vacuum.
7. Dolphins communicate by making
   a. low-frequency sounds.
   b. high-intensity sounds.
   c. high-frequency sounds.
   d. low-intensity sounds.
8. Music consists of rhythm, melody, and
   a. intensity.
   b. harmony.
   c. percussion.
   d. frequency.
9. Before your eardrum vibrates, the sound wave has to travel through the
   a. anvil.
   b. ear canal.
   c. inner ear.
   d. cochlea.
10. A sound wave consists of compressions and
    a. overtones.
    b. fundamental tones.
    c. rarefactions.
    d. frequencies.

TRUE/FALSE Write *true* if the statement is true. If the statement is false, change the underlined term to make the statement true.

11. Sound cannot travel through a vacuum.
12. A high-intensity sound wave has a lot of energy.
13. Sound waves cause the ear canal to vibrate.
14. An echo is a sound wave that has been made louder.
15. One hertz is equal to one compression per second.
16. A drum is an example of a percussion instrument.
17. The beat of the music is the melody.
18. There are three main types of musical instruments.
19. Vibrations from the eardrum enter the outer ear.
20. Humans can hear sounds between 20 Hz and 20,000 Hz.

Content Challenges TEST PREP

WRITTEN RESPONSE **Answer each of the following questions in complete sentences.**

1. **HYPOTHESIZE:** Why does sound travel faster in a solid than in a gas?
2. **PREDICT:** What happens after a sound wave makes the eardrum vibrate?
3. **INFER:** What does a vibrating guitar string do to the air around it?
4. **EXPLAIN:** Which sound wave has a larger amplitude, a 20-dB sound or a 50-dB sound? How do you know?
5. **PREDICT:** What might happen to your sense of hearing if you were exposed to loud noises over a long period of time?

INTERPRETING VISUALS **Use Figure 3-31 to answer the following questions.**

6. When a sound is heard, what part of the ear vibrates first?
7. Which parts make up the inner ear?
8. Where is the energy of a sound wave converted to electrical impulses?
9. Why do you think the bones in the middle ear are called the hammer, the anvil, and the stirrup?
10. Cupping your hand to your ear helps you to hear sounds more clearly. Why?

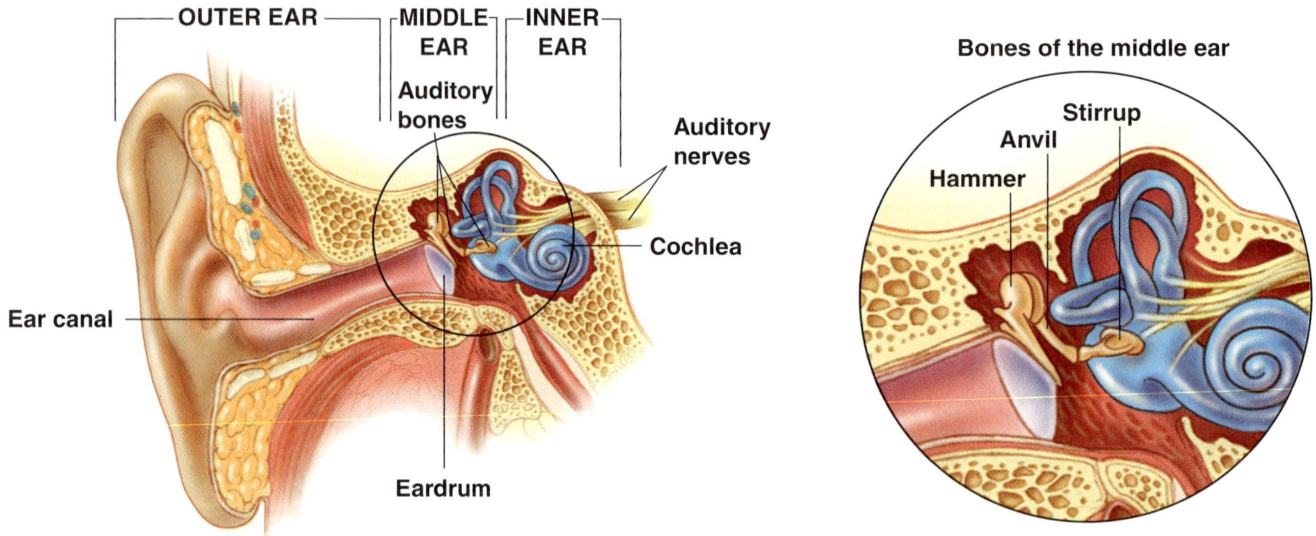

▲ **Figure 3-31** The human ear

Chapter 4 Light

▲ **Figure 4-1** A bundle of optical fibers

Light is a form of energy. Much of our information comes to us in the form of light images. Light can travel through empty space. It can also travel through other materials, such as glass, water, and air. Many technologies that use light, such as fiber optics, have been developed. In fiber optics, light is carried through thin cables of a glasslike material. Fiber optics are used in many areas of life, particularly communications.

►Name something you use daily that works by fiber optics.

Contents

4-1 What is the electromagnetic spectrum?

Objective

Identify the parts of the electromagnetic spectrum.

Key Terms

electromagnetic (ee-lehk-troh-mag-NEHT-ihk) **spectrum:** range of electromagnetic waves

visible spectrum: seven colors that make up white light

Electromagnetic Waves Light is a form of energy. Sometimes light behaves as if it is made up of tiny particles of energy called photons. Unless disturbed, photons travel in straight lines. Other times, light behaves as an electromagnetic wave. Unlike mechanical waves, electromagnetic waves can travel through empty space.

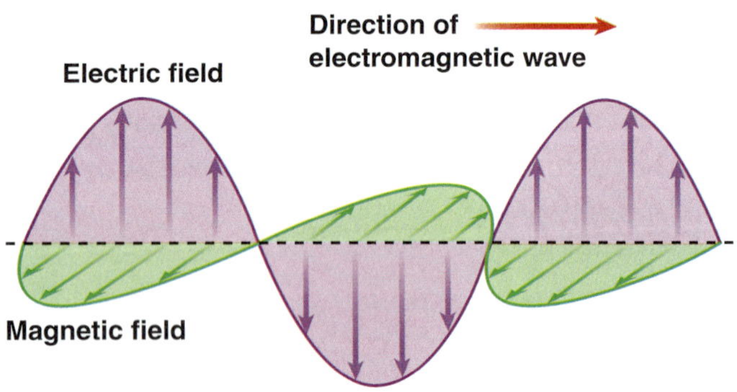

▲ **Figure 4-2** Electromagnetic waves are made up of two transverse waves.

Electromagnetic waves have both electrical properties and magnetic properties. These waves are made up of two transverse waves that vibrate at right angles to each other.

1 COMPARE: How are electromagnetic waves different from mechanical waves?

The Electromagnetic Spectrum Light that you can see is called visible light, or white light. Visible light waves make up a small part of a much larger band of waves called the **electromagnetic spectrum.** Figure 4-3 shows the different types of waves that make up the electromagnetic spectrum. As the figure shows, the energy of a wave is related to its wavelength and its frequency.

- **Radio waves** Radio waves have the longest wavelength and the lowest frequency. They are used for radio, television, and radar signals.
- **Microwaves** Microwaves are used for wireless communications and microwave ovens.
- **Infrared rays** All objects give off infrared rays. You cannot see infrared rays, but you can feel them as heat.
- **Visible spectrum** The only part of the electromagnetic spectrum that can be seen by the human eye is the **visible spectrum.** It is made up of the colors red, orange, yellow, green, blue, indigo, and violet.

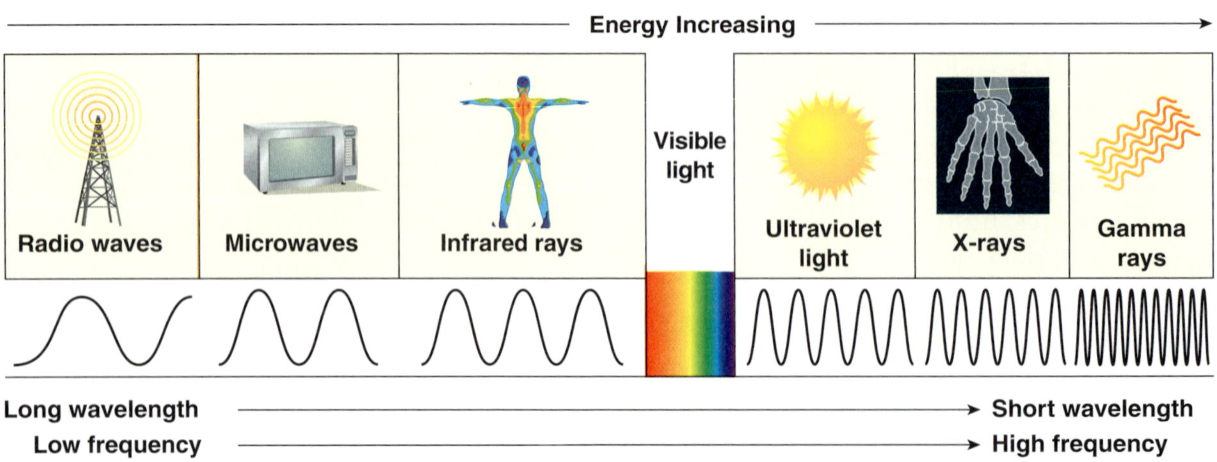

▲ **Figure 4-3** The electromagnetic spectrum is made up of different wavelengths and frequencies.

- **Ultraviolet rays** Ultraviolet rays can cause sunburn. Ultraviolet light is used to kill bacteria.
- **X-rays** X-rays have a very short wavelength and a high frequency. X-rays are used in medicine to form images of bones and internal organs.
- **Gamma rays** Gamma rays have a shorter wavelength and a higher frequency than X-rays do. They are given off during nuclear reactions.

2 **LIST:** What are the electromagnetic waves that make up the electromagnetic spectrum?

CHECKING CONCEPTS

1. Waves with the longest wavelength and the lowest frequency are __________.
2. Waves usually given off during nuclear reactions are __________.
3. Only those waves in the __________ can be seen by the human eye.
4. Waves with wavelengths slightly longer than visible light are __________.
5. Rays usually felt as heat are __________.

THINKING CRITICALLY

6. **HYPOTHESIZE:** Microwaves are high-frequency radio waves. Microwave ovens can be used to cook food quickly. The microwaves are absorbed by the food. Why do you think that food in a microwave oven gets hot, but a glass dish does not?

HEALTH AND SAFETY TIP

High-frequency electromagnetic waves are dangerous. Too much exposure to X-rays is harmful. Find out why X-ray technicians wear lead aprons or stand behind a lead screen when taking X-ray pictures. Why are pregnant women advised not to have X-rays taken?

Science and Technology

CELL PHONES

Cellular or cell phones have helped change the way we communicate. These phones are small radio transmitters and receivers. One of hundreds of antennae receives radio signals when you make a call. These radio signals are transmitted on a radio frequency to a local cell site. A mobile telephone switching office (MTSO) picks up the signals and sends them through regular wired phones. The MTSO tracks and transfers your calls as you move from area to area. It reverses when you get a call.

Two kinds of wireless phone networks are analog and digital. In analog, voices are carried on radio waves with limitations. This service is easy to steal, and busy signals and static are common. Conversations can be picked up on radio scanners.

▲ **Figure 4-4** Cell phones receive and transmit radio waves.

Digital service is newer. Voices convert to computer language and then change back to voices on the receiving end. Busy signals and static are less common. Your conversation can not be heard on a scanner. This service is hard to steal. The newest technology is PCS (personal communication service). Computer language is transmitted on a range of frequencies that are always changing. This technology allows you to access the Internet, and have paging and text messaging capabilities.

Thinking Critically How have cell phones changed the way we communicate?

4-2 What is light?

INVESTIGATE

Observing How Light Travels

HANDS-ON ACTIVITY

1. Stack three index cards one on top of the other and make a hole in the center.
2. Fold one edge of each card vertically. The folded edge should be about 4 cm wide.
3. Stand the cards in a row as shown. Line them up so that you can see through the holes.
4. Have a classmate shine a flashlight through the holes in the cards. Move the center card a few centimeters to the left or right. Observe what happens.

STEP 4

THINK ABOUT IT: Can you see the light when the cards are lined up? Can you still see the light when you move the center card? What do these results tell you about the way light travels?

Objective

Describe light as a form of electromagnetic energy.

Key Terms

light: form of electromagnetic energy made up of streams of photons

photon (FOH-tahn)**:** tiny bundle of energy

ray: straight line that shows the direction of light

Light Energy When you sit in bright sunlight, you can feel your skin get warm. Objects warm up in sunlight. **Light** is a form of electromagnetic energy. Light energy can be changed into heat, electricity, and other forms of energy.

1 **CLASSIFY:** What is light?

Particles of Light Light sometimes behaves as if it is made up of small bundles of energy called **photons**. Photons are such small particles that a single photon cannot be seen. A beam of light is made up of a stream of many photons. Each photon carries a certain amount of energy. Some photons have more energy than others.

2 **DESCRIBE:** Of what is a light beam made?

Rays of Light When you turn on a flashlight, you see a beam of light. The beam of light from the flashlight looks like a straight line. Light travels in straight lines. A **ray** of light is a straight line that shows the direction of a light beam. A ray of light will continue to travel in a straight line unless its direction is changed. There are many ways to model light. Scientists model light in different ways, depending on the question they want to answer.

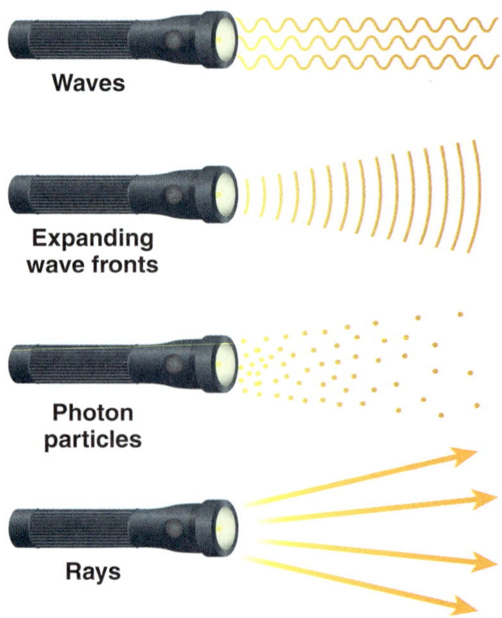

▲ **Figure 4-5** Four models of light

3 **DESCRIBE:** What does a ray of light look like?

CHECKING CONCEPTS

1. You cannot see individual __________ of light because they are very small.
2. Light travels in __________ lines.
3. A stream of photons makes up a __________ beam.
4. Light is a form of __________ energy.
5. A __________ of light shows the direction of a light beam.

THINKING CRITICALLY

6. **INFER:** Turn on a lamp and hold your hand near the light bulb. Do you feel heat? What is the source of the heat?
7. **EXPLAIN:** Give two common examples to show that light travels in straight lines.

Web InfoSearch

Light Years: Speed of Light The speed of light is 300,000 km/s. Light takes less than two seconds to reach Earth from the Moon. The distance that light travels in one year is called a light year. The nearest star is 4.2 light years away from Earth. Light from this star takes 4.2 years to reach Earth.

SEARCH: Use the Internet to find out more about light years. How are light years used? What is another measurement of distance in space? Start your search at www.conceptsandchallenges.com. Some key search words are **light years, measuring distance stars,** and **stellar distances.**

How Do They Know That?

PHOTONS AND THE THEORY OF LIGHT

Research for hundreds of years supports the idea that light is made up of particles of energy called photons.

In the late 1600s, Sir Isaac Newton studied how light acts as it meets objects in its path. Newton noticed that light could not travel around objects like other waves could. For example, sound can travel around a corner, but light cannot. He suggested that light was made up of tiny streams of energy particles.

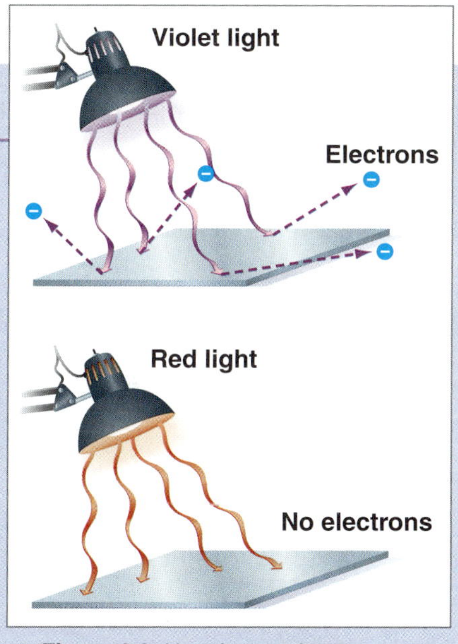

▲ **Figure 4-6** Hertz's experiment

In 1887, Heinrich Hertz found that some frequencies of light, such as violet, would cause metals to give off their electrons. When he used red light, no matter how strong, nothing happened. This experiment showed that the particles in violet light were more energetic than those in red light. This is called the photoelectric effect.

In 1900, Max Planck published a theory on electromagnetic radiation. Albert Einstein used this theory to help explain his theory on photons. He said that a photon's energy depends on the frequency of the light. This theory won Einstein the Nobel Prize.

The photoelectric effect has many everyday uses. Light meters in cameras, soundtracks in films, and automatic street lights all work because of the photoelectric effect. The photons in a light source make an electric current that controls these light-sensitive machines.

Thinking Critically Why do you think it is important to understand the nature of light?

4-3 How do light waves travel?

Objective

Describe how light travels as transverse waves.

Key Term

transverse (trans-VURS) **wave:** wave in which the particles of the medium move up and down at right angles to the direction of the wave motion

Light Waves Light is made up of streams of photons. However, light also behaves like a wave. Light is a type of electromagnetic wave. Electromagnetic waves are different from sound waves. Sound is a longitudinal wave. A sound wave needs a medium in which to travel. Sound cannot be heard in a vacuum, or empty space.

Light waves are different from sound waves in two ways. Light travels in transverse waves. In transverse waves, the particles move up and down at right angles to the direction of wave motions. Also, light waves do not need a medium in which to travel. Light can travel through a vacuum.

1 CONTRAST: How do light waves differ from sound waves?

Photons and Light Waves Some experiments with light show that it is made up of photons. Other experiments show that light acts as a wave. Scientists have learned that some waves act as if they are made up of particles.

2 DESCRIBE: How do light waves act?

Properties of Light Waves Like all waves, light waves have four properties. They are speed, wavelength, frequency, and amplitude.

- The **speed** of light is 300,000 km/s in a vacuum. Light and all other electromagnetic waves travel at this speed. The speed of light in a vacuum is the fastest possible speed.
- The **wavelength** of light is the distance from the crest or trough of one wave to the crest or trough of the next wave.
- The number of light waves that pass by a point each second is the **frequency.**
- The **amplitude** is the height of a wave. A bright light has a greater amplitude than a dim light does.

3 LIST: What are the four properties of a light wave?

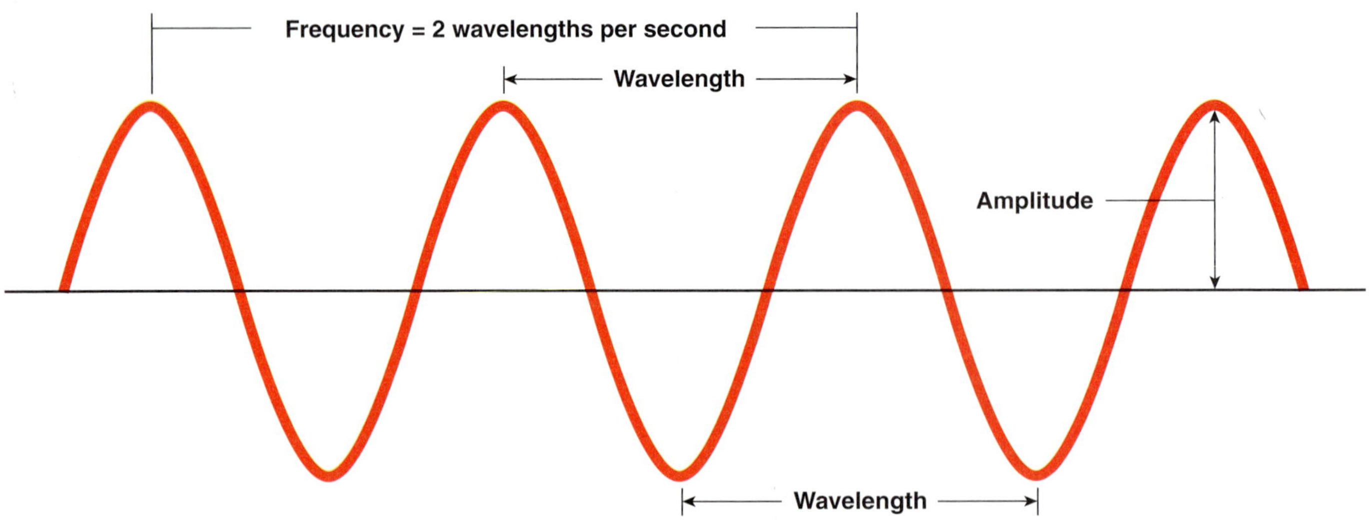

▲ **Figure 4-7** Parts of a transverse wave

CHECKING CONCEPTS

1. In a __________ wave, the particles of the medium move at right angles to the direction of the wave motion.
2. A light wave is a moving stream of __________.
3. The properties of light waves include speed, __________, frequency, and amplitude.
4. Light waves travel fastest in a __________.
5. Light waves do not need a __________ in which to travel.

THINKING CRITICALLY

6. **CLASSIFY:** Identify each of the following as a sound wave or a light wave.
   a. longitudinal wave
   b. does not travel through a vacuum
   c. waves seem to be made up of particles
   d. transverse waves
   e. can travel in a vacuum
7. **RELATE:** Light waves with very high frequencies are called ultraviolet light. Ultraviolet light cannot be seen by the human eye. How is ultraviolet light similar to ultrasonic sound?

INTERPRETING VISUALS

Look at Figure 4-8. Compare the two waves by discussing wavelength, frequency, amplitude, and speed.

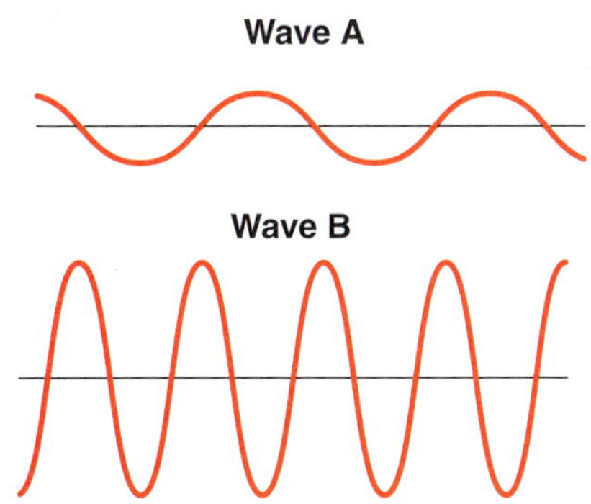

▲ **Figure 4-8** Compare the waves.

Real-Life Science

PHOTOGRAPHY

Photography is a popular activity that can teach you a lot about light. The word *photography* means "writing with light." To take a photograph, you need a camera, film, and a good source of light.

A camera has a lens and an opening for light to enter. This opening is called an aperture. When you take a picture, the aperture is open for only a short period of time. The light goes through the lens and onto the film. Film is very sensitive to light. Too much light will cause the photograph to be too bright, or overexposed. If there is not enough light, the photograph will be too dark, or underexposed. This is because the chemical on the film changes whenever a photon hits it.

Many schools have amateur photography clubs. Joining a photography club is a good way to learn more about taking photographs.

▲ **Figure 4-9** The varied light made parts of this photo brighter and darker.

Thinking Critically How is film underexposed or overexposed?

4-4 What are sources of light?

Objectives

Differentiate between luminous and illuminated objects. Describe what happens when light strikes different materials.

Key Terms

luminous (LOO-muh-nuhs) **object:** object that gives off its own light

illuminated (ih-LOO-muh-nayt-uhd) **object:** object that reflects light

opaque (oh-PAYK)**:** material that blocks light

transparent (trans-PEHR-uhnt)**:** material that transmits light easily

translucent (trans-LOO-suhnt)**:** material that transmits some light

Luminous Objects A flashlight, a candle, and a light bulb are sources of light. Objects that give off their own light are called **luminous objects.** The Sun is a luminous object. It is the source of light for Earth and other planets in the solar system.

1 **DEFINE:** What is a luminous object?

Illuminated Objects What happens when the light from a luminous object strikes another object? The light may be reflected. An object that reflects light is called an **illuminated object.**

The Moon is an illuminated object. The Moon is not the source of its own light. Sunlight strikes the Moon and is reflected from its surface. You see the Moon by reflected sunlight.

2 **IDENTIFY:** What is an object that reflects light called?

Opaque, Transparent, and Translucent Materials When light shines on a material, one of three things can happen. The material may block the light and form a shadow. Material that blocks light is called an **opaque** object. When you place your hand in front of a light source, you will see a shadow. Your hand is opaque.

A material may allow light to pass through unchanged. This kind of material is called **transparent.** A sheet of clear glass allows light and images to pass through. The glass is transparent.

A material may allow some of the light to pass through. This kind of material is called **translucent.** A stained-glass window will allow some light to pass through, but no details of images are visible. The stained glass is translucent.

3 **NAME:** What are three types of materials that affect light differently?

Opaque **Translucent** **Transparent**

◀ **Figure 4-10** You cannot see the stem of the flower through the opaque vase. A blurry view of the stem is seen through the translucent vase. The entire stem can clearly be seen through the transparent vase.

CHECKING CONCEPTS

1. When you shine a flashlight on objects in a dark room, the objects that you see are ___________ objects.
2. An object that blocks light and casts a shadow is an ___________ object.
3. The Moon is an ___________ object.
4. A sheet of clear glass is ___________ because it allows light to pass through easily.
5. A sheet of stained glass is ___________ because it only allows some light to pass through.
6. An illuminated object ___________ the light from another source.
7. The Sun is a ___________ because it gives off light.

THINKING CRITICALLY

8. **INFER:** You cannot see the Sun on a cloudy day even though sunlight can still illuminate objects. Are clouds transparent or translucent?
9. **CLASSIFY:** Which of the following objects are luminous and which are illuminated?
   - **a.** the flame of a candle
   - **b.** Earth
   - **c.** a star
   - **d.** a campfire
   - **e.** a sheet of paper

DESIGNING AN EXPERIMENT

Design an experiment to solve the following problem. Include a hypothesis, variables, a procedure, and a type of data to study.

PROBLEM: Do transparent objects allow more light to pass through than translucent objects do?

Integrating Earth Science

Topic: lunar eclipse

ASTRONOMICAL SHADOWS

Opaque objects cast shadows by blocking the light shining on them. Your body is opaque. It blocks the Sun's light. You have probably seen your shadow on a sunny afternoon.

During a lunar eclipse, Earth blocks the light from the Sun. The Moon passes through Earth's shadow. Like most shadows, Earth's shadow consists of both an umbra (UM-bruh) and a penumbra (pih-NUM-bruh). The umbra is a sharp, black inner part of a shadow. The penumbra is the gray, outer part of the shadow.

When a point source shines directly on an object, the shadow will only have an umbra. A spotlight is an example of a point source of light. Most light comes from sources of light that spread out. A fluorescent light is an example of a source of light that spreads out.

Thinking Critically Why do most shadows have penumbras?

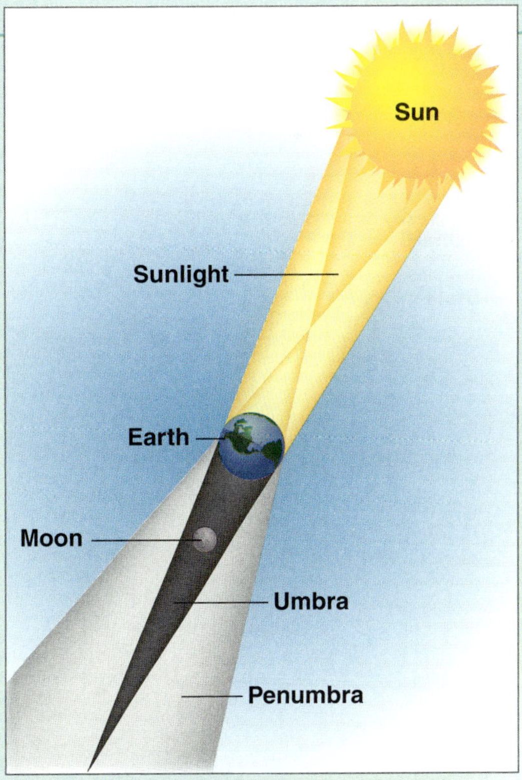

▲ **Figure 4-11** During a lunar eclipse, Earth hides the Moon from the Sun.

4-5 What is photosynthesis?

Objective

Explain how plants use the energy of the Sun to make food.

Key Terms

photosynthesis (foht-oh-SIHN-thuh-sihs): process by which plants use energy from the Sun to make food

chlorophyll (KLAWR-uh-fihl): substance in plants that absorbs the Sun's light and gives plants their green color

chloroplast: part of green plant cells where photosynthesis takes place

Light and Energy Any form of energy can be converted to another form of energy. Light energy can be changed into heat energy, electrical energy, chemical energy, or other forms of energy. Plants convert the electromagnetic energy of sunlight into the chemical energy stored in food. The plants use the chemical energy to make their own food.

1 **INFER:** Why do plants need sunlight?

Photosynthesis The process by which a plant uses sunlight to make food is called **photosynthesis.** The leaves of a plant absorb the most sunlight. The flat part of a leaf always tries to face the Sun. As the Sun moves across the sky, the leaves respond by moving to continue facing the Sun. The food that plants make is sugar. Plants can change this sugar into starch, fats, and proteins. These nutrients are stored in the plants as chemical energy. This energy can be used at a later time.

Photosynthesis is a chemical process. During this process, water, carbon dioxide and energy from the Sun are used to make the sugar. Roots absorb the needed water from the soil. Veins carry the water up the stem and into the leaves. Carbon dioxide enters the plant through openings called stomata. Sunlight supplies the energy the plant needs to make the sugar. During photosynthesis, oxygen is given off as a by-product. The formula for photosynthesis is shown below.

$$6CO_2 + 6H_2O \xrightarrow[\text{Sunlight}]{\text{Energy}} C_6H_{12}O_6 + 6O_2$$

Carbon dioxide + Water → Sugar + Oxygen

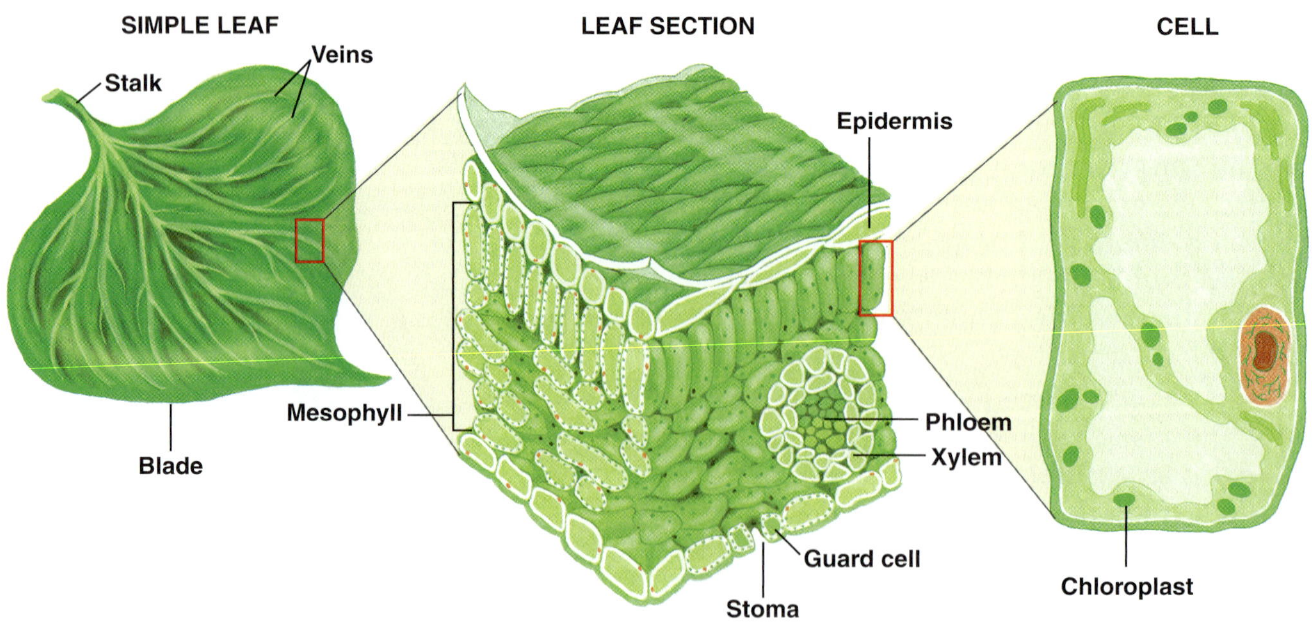

▲ **Figure 4-12** Photosynthesis takes place in the chloroplasts of a leaf cell.

▲ **Figure 4-13** A rain-forest ecosystem

Most organisms depend on plants and photosynthesis for survival. When animals eat plants, they take in food that has stored chemical energy. Animals that only eat other animals still depend on plants because the animals that they eat, eat plants. Plants are an important part of most ecosystems, such as the rain forest shown in Figure 4-13.

2 IDENTIFY: What is the process by which plants use sunlight to make food?

Chlorophyll The substance in a leaf that absorbs sunlight is called **chlorophyll.** Chlorophyll is a chemical pigment, or coloring, needed for photosynthesis. It absorbs all the colors of the visible spectrum except green, which it reflects. As a result, leaves look green. Without chlorophyll, photosynthesis could not take place. Wherever there is chlorophyll, photosynthesis can occur.

The chlorophyll is located inside special parts of plant cells. The cell parts that contain chlorophyll are called **chloroplasts.** The cells in the upper layer of the leaf contain many chloroplasts. Therefore, most photosynthesis takes place in this layer.

3 PREDICT: What color would leaves appear if chlorophyll reflected only blue light?

CHECKING CONCEPTS

1. In photosynthesis, electromagnetic energy is turned into __________ energy.
2. The food that a plant makes is __________.
3. Food is a form of __________ energy.
4. The substance that gives plants their green color is __________.
5. A plant's __________ always face the Sun in order to get the most amount of sunlight.
6. Chlorophyll absorbs all the Sun's light except __________ light.

THINKING CRITICALLY

7. INFER: Plants always make more food than they need. The extra food is stored in the plant. What happens when an animal eats a plant?
8. HYPOTHESIZE: What happens to the energy that plants do not store as food?

Web InfoSearch

Photoelectric Cells A process that can be compared to photosynthesis is used to produce electricity from solar energy. Sunlight is absorbed by photoelectric cells. Photoelectric cells convert sunlight into electrical energy. The electrical energy is stored in batteries. This process is similar to what a plant does, except that the plant stores the energy as food. Many photoelectric cells are put together to make solar panels. Solar panels are used on homes, office buildings, and satellites.

SEARCH: Use the Internet to write a report on how a solar panel works. Start your search at www.conceptsandchallenges.com. Some key search words are **solar panels, photoelectric cells, photovoltaic panels,** and **solar power.**

4-6 How do lenses refract light?

Objective

Compare and contrast how light rays are bent by a concave lens and a convex lens.

Key Terms

lens: transparent material that bends light

real image: image that can be projected onto a screen

virtual image: image that cannot be projected onto a screen

convex lens: lens that curves outward

concave lens: lens that curves inward

Refraction of Light Light rays travel in straight lines. When light moves from one medium into another at an angle other than 90°, the direction of the light changes. Suppose a beam of light passes from air into another material, such as glass. The path of light will be bent, or refracted. Light rays passing through the lenses in Figures 4-14 and 4-15 are refracted.

1 **PREDICT:** What happens to a beam of light when it passes from air into water?

Lenses A **lens** is a transparent material that bends, or refracts, light. All lenses have either one or two curved surfaces. There are two main types of lenses. They are convex and concave lenses.

2 **HYPOTHESIZE:** Why is a lens made of a transparent material?

Real and Virtual Images A lens can form a real image. A **real image** is an image that can be projected onto a screen.

The image you see when you look into a mirror is a **virtual image.** It is always right-side up. This image is called virtual because the image does not really exist, even though you see it. A virtual image cannot be projected onto a screen.

3 **DEFINE:** What is a real image?

Convex Lenses A **convex lens** curves outward. Figure 4-14 shows a convex lens. Light that passes through a convex lens is bent inward. When a convex lens refracts light, the light rays are brought together at a point called the focal point. The distance between the lens and the focal point is the focal length.

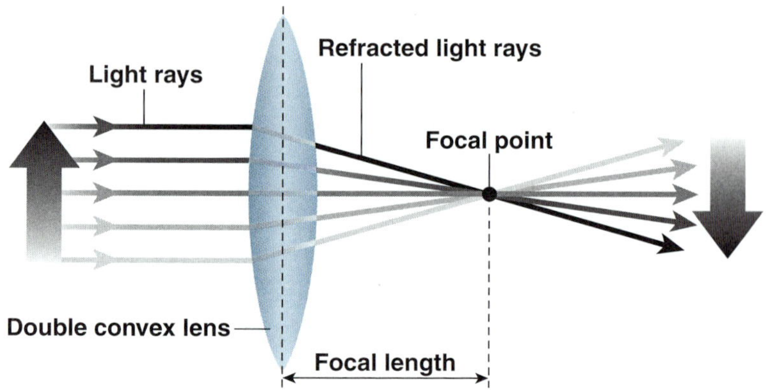

▲ **Figure 4-14** A double convex lens forms a real image.

Light is refracted as it enters the lens. It also is refracted as it leaves the lens. The amount of refraction depends on how curved the lens is.

4 **IDENTIFY:** What is the focal length of a convex lens?

Concave Lenses A **concave lens** curves inward. Light that passes through a concave lens is bent away from the lens. The light rays are spread apart. Figure 4-15 shows a concave lens.

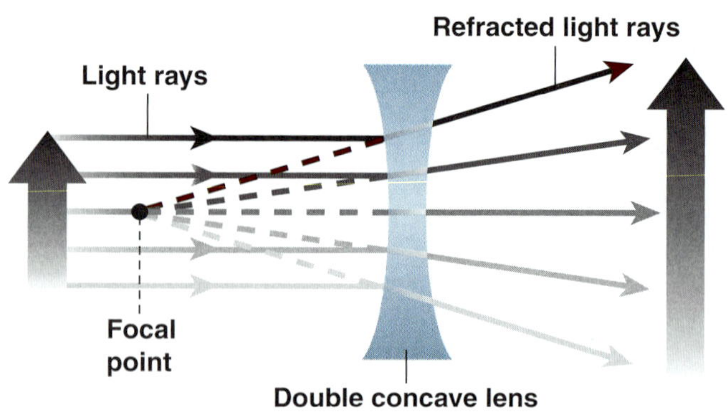

▲ **Figure 4-15** A double concave lens forms a virtual image.

5 **DESCRIBE:** What happens to light rays as they pass through a concave lens?

CHECKING CONCEPTS

1. A __________ image cannot be projected onto a screen.
2. A __________ lens curves inward.
3. A __________ can be projected onto a screen.
4. A __________ lens curves outward.
5. The __________ is where rays of light are brought together by a convex lens.
6. The __________ is the distance from the focal point to the lens.
7. A __________ image will be right-side up.

THINKING CRITICALLY

8. **COMPARE:** Explain the difference between a real image and a virtual image.
9. **ANALYZE:** How are light rays brought together after passing through a convex lens?

Web InfoSearch

Telescopes A telescope is a device that allows you to see objects at great distances. From the giant Hubble Space Telescope, which orbits Earth, to the personal telescope used at home or at school, telescopes come in many sizes and strengths.

SEARCH: Use the Internet to find out more about telescopes. How do they work? Why do some telescopes show an upside-down image? Start your search at www.conceptsandchallenges.com. Some key search words are **refractor telescopes, how telescopes work,** and **telescopes' refractor lenses.**

People in Science

ASTRONOMERS

Astronomy is the study of stars, planets, and the solar system. Astronomers are the scientists who study how these objects originated, what they are made of, and how and where they move. Astronomers may choose from many different jobs. Some study the formation of galaxies, solar astronomy, the origins of stars, or planetary science. Some astronomers design programs to observe the universe; others observe just our solar system.

Astronomers have a very hard job because they cannot see, touch, smell, weigh, or perform experiments on the objects they are studying. Instead they need to observe the light given off or reflected by objects. The motion of objects is also measured. In order to become an astronomer, a solid background in physics is necessary. You also need to know how to work a computer. Good math, observation, reasoning, and communication skills are necessary.

▲ **Figure 4-16** This is the telescope and data recording area of the Lowell Telescope at Perth Observatory in Australia.

Astronomers do not spend all day looking through a telescope. They study the data recorded by a telescope and sent to a computer. Then, they have to put that information into readable form. Astronomers also give lectures, teach, and write articles.

Thinking Critically Why is an astronomer's job difficult?

4-7 How do you see?

INVESTIGATE

How Does the Eye React to Light?

HANDS-ON ACTIVITY

1. Work with a partner. You need a penlight, a sheet of paper, and a pencil.
2. Stand face-to-face.
3. Darken the room. Record what you observe about each other's eyes.
4. Now, shine the penlight into the eye of your partner. Record what you observe in your partner's eye.
5. Alternate roles so you each get a chance to observe what happens.

THINK ABOUT IT: In dim light, your iris contracts and the pupil gets bigger. What happens in bright light?

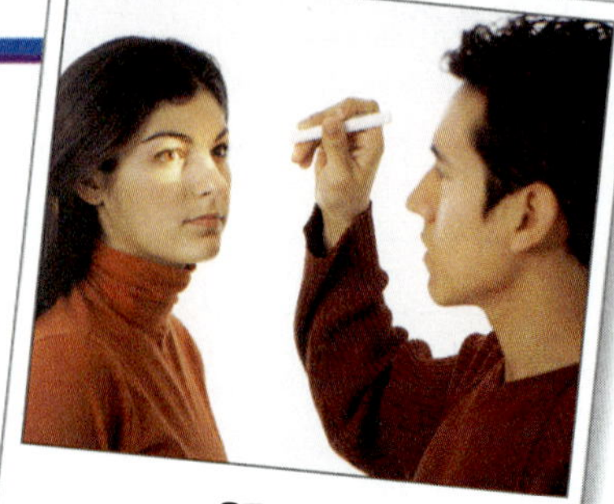
STEP 4

Objectives

Identify the parts of the eye. Describe how the eye senses light.

Key Terms

eye: sense organ that detects light

sight: one of the five human senses

image: picture formed by the eye

Observing Light You see an object when light from the object enters your eye. The **eye** is the human organ that detects light. **Sight** is one of the five human senses.

1 RELATE: How is the eye related to sight?

Parts of the Eye The eye is made up of parts that work together to help you see. Look at Figure 4-17. The front of the eye is covered with a clear layer called the cornea. The cornea is a protective layer that keeps dirt and bacteria from damaging the inner eye. The cornea covers the pupil. The pupil is the opening through which light enters the eye.

The amount of light that enters the eye is controlled by the iris. The iris is a muscle that contracts or expands, depending on the amount of light available. When there is too much light, the iris contracts. When there is not enough light, the iris expands.

After light enters the eye, it is focused by the lens. It is the clear part behind the pupil. The lens can change shape so that the eye can concentrate on an object. The lens works like the autofocus on a camera.

2 NAME: What are the parts of the eye involved in sight?

The Retina Light entering the eye is focused by the lens onto the retina. The lens, a convex lens, focuses light on the retina, producing an image. An **image** is a picture that is formed by the eye.

The retina converts the image into electrical impulses. The impulses travel along the optic nerve to the brain. The brain interprets the electrical impulses as a right-side-up image.

3 RELATE: How do the lens and the retina work together?

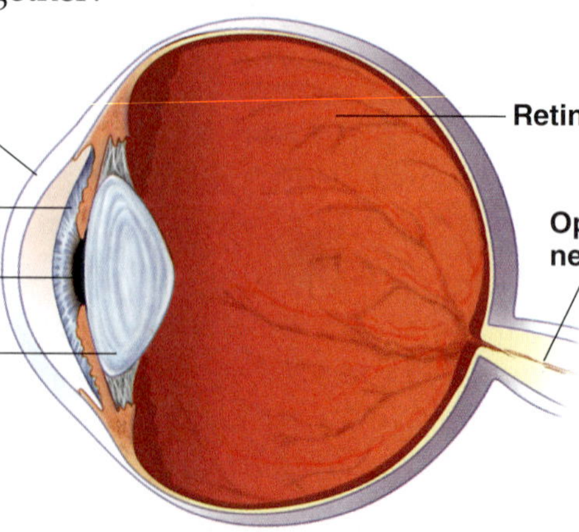

▲ **Figure 4-17** The eye, organ of sight

CHECKING CONCEPTS

1. Electrical impulses travel along the __________ to the brain.
2. The __________ is the opening that lets light into the eye.
3. The __________ controls the amount of light that can enter the eye.
4. The __________ is a clear protective layer over the eye.
5. The eye is a sense organ that detects __________.
6. An image formed on the __________ is upside down.

THINKING CRITICALLY

7. **INFER:** What energy changes take place when you see an image?
8. **SYNTHESIZE:** Suppose you walk into a dark room after being out in bright sunlight. Several minutes pass before you can see objects in the room clearly. What happened to your eyes during that time?

INTERPRETING VISUALS

Use Figure 4-18 to answer the following questions.

9. What type of corrective lens is used to correct nearsightedness?
10. What type of corrective lens is used to correct farsightedness?

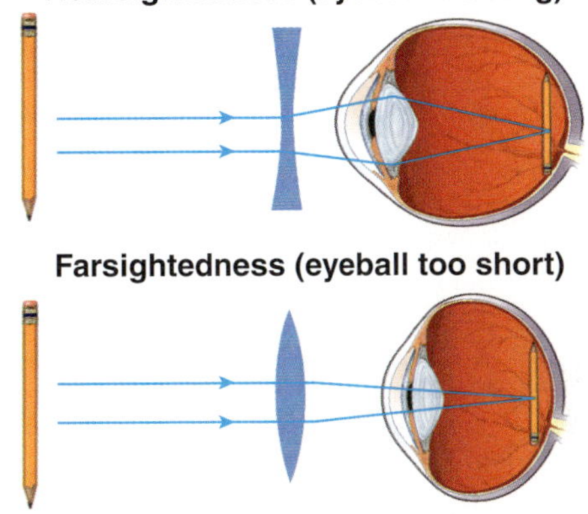

▲ **Figure 4-18** Identify the lenses.

Science and Technology

LASIK EYE SURGERY

Laser is an acronym for light amplification by the stimulated emission of radiation. Lasik (LAY-sik) is an acronym for laser in situ keratomileusis. Lasik eye surgery corrects some vision problems using lasers.

Lasik has been practiced in the United States since 1996. It was invented in Colombia by Drs. Virgilio Galvis and Luis Ruiz. The laser used and approved in the United States is called the VISX. It can correct nearsightedness and some other eye problems. During lasik eye surgery, a flap of the cornea is cut and folded over. A laser beam then takes away some of the tissue from the inside of the cornea. A metal ring is used with an infrared tracking device to make sure that the laser points to the right spot. The laser reshapes that part of the cornea. Then, the flap is put back.

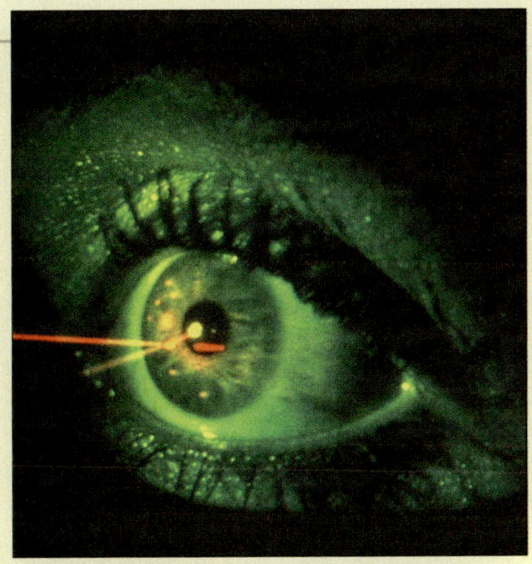

▲ **Figure 4-19** Laser surgery corrects vision by reshaping the cornea.

There are benefits if you can have the surgery. It is same-day surgery. You have it done in the doctor's office and can leave after a short stay. However, not everyone with poor vision can be helped by this surgery. Also, like any surgery, lasik surgery does have some risk involved, such as infection. It is very important to seek medical advice before allowing any surgery to be done.

Thinking Critically Why do you think recovery from lasik surgery is so quick?

LAB ACTIVITY
Comparing Convex and Concave Lenses

Materials

- Convex lens (10-cm focal length),
- Concave lens (10-cm focal length),
- Flashlight,
- Metric ruler,
- Dark construction paper or cardboard,
- 1 sheet of white paper,
- Notebook paper,
- Tape, scissors, pencil
- Hardcover book

BACKGROUND

A lens is an object that bends, or refracts, light. When light passes through a lens, an image is created. Images can be real or virtual. A real image will be inverted, or upside down. A virtual image will be right-side up. Real images can be projected; virtual ones cannot. There are two types of lenses, convex and concave. Convex lenses curve outward in the center. Concave lenses curve inward in the center.

PURPOSE

In this activity, you will compare the images formed by convex and concave lenses.

PROCEDURE

1. Copy the chart shown in Figure 4-21 onto a sheet of paper.
2. Write your name on a sheet of notebook paper. Write it as you normally write it—not too big and not too small.
3. Hold the convex lens over your name and look through it. Slowly lift the lens toward your eye. Observe your name through the lens. You may have to move the lens back and forth to focus the image of your name. Record your observations on your chart.
4. Repeat Step 3 using the concave lens.
5. Create a "projection screen" as shown in the Step 5 photo. Tape a sheet of white paper onto a large hardcover book. Then, stand the book upright so that the "screen" is at one end of your desk or table.
6. Now, make a "projector." Lay the construction paper flat on a table. Place the head of the flashlight onto the paper and trace it to make a circle. Cut out the circle. Fold the circle in half. In the center, draw

▲ **STEP 3** Hold the lens over your name.

▲ **STEP 5** Set up a projection screen.

a triangle connected to a rectangle along the folded edge of the paper, as shown in Figure 4-20. Now, cut out the shape and unfold the paper. You should have an arrow-shaped hole in the middle of the circle. Tape the circle over the head of the flashlight. Be careful not to put tape over the cutout arrow.

◀ **Figure 4-20** Draw and cut out this shape.

7. Hold the flashlight about 50 cm from the "screen." Turn on the flashlight and point it at the screen. Hold the convex lens in front of the lighted flashlight. Move the lens back and forth to focus the image of the arrow on the screen. You may have a partner hold the light steady while you focus. You also may need to turn the lens slightly until you see the arrow clearly. Record your observations.

8. Now, repeat Step 7 using the concave lens. Record your observations. Remember, only real images can be projected.

▲ **STEP 7** Focus the lens so you see an image on the screen.

Comparison of Convex and Concave Lenses

	Eye	Flashlight
Convex		
Concave		

▲ **Figure 4-21** Copy this chart and use it to record your observations.

CONCLUSIONS

1. **OBSERVE:** What did the convex lens do to your name?
2. **OBSERVE:** What did the concave lens do to your name?
3. **ANALYZE:** Which lens allowed you to see real images? How do you know?
4. **INFER:** Which type of lens do you think would be used in a magnifying glass?
5. **SYNTHESIZE:** How are different lenses used to correct people's vision?

4-8 How do mirrors reflect light?

Objective

Describe how mirrors form clear images.

Key Terms

regular reflection: reflection that forms a clear image

diffuse reflection: reflection that forms a fuzzy image

plane mirror: smooth surface that reflects light and forms images

Reflected Light Light not completely absorbed or transmitted by an object will be reflected. The type of surface affects the type of reflection. There are two types of reflection. They are regular reflection and diffuse reflection.

When a beam of light strikes a smooth surface, the light is reflected from that surface at the same angle. A **regular reflection** forms a clear image. For example, the light reflected from the surface of a still pond will form clear images of the objects around the pond. A regular reflection only occurs when a surface is completely smooth and even, as shown in Figure 4-22.

▲ **Figure 4-22** Regular reflection

▲ **Figure 4-23** Diffuse reflection

When light strikes a rough surface, it will be reflected at different angles. The reflected light from a **diffuse reflection** forms a fuzzy image because light is scattered in many directions. On a windy day, the surface of a pond is not smooth. The light reflected from the surface of the pond will form fuzzy images as shown in Figure 4-23. It is by diffuse reflection that we are able to see nonluminous objects.

1 IDENTIFY: What are the two types of reflection called?

Plane Mirrors Mirrors form regular reflections. A **plane mirror** is a smooth surface that reflects light and forms images. The light from an object striking a mirror at an angle will be reflected from the mirror at the same angle. Figure 4-24 shows a baseball reflected in a mirror at a 70° angle. This is the angle of incidence. The reflected beam bounces off the mirror at a 70° angle. This is the angle of reflection. According to the law of reflection, the angle of incidence is equal to the angle of reflection. The image formed by the mirror is the same distance behind the reflecting surface as the object is in front of it. Figure 4-24 shows that distances *A* and *B* are equal.

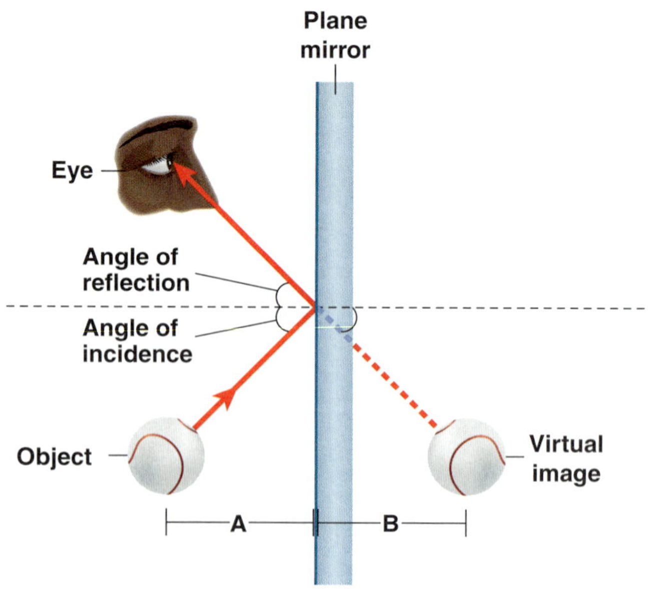

▲ **Figure 4-24** Mirror image

2 INFER: Why does a mirror form a clear image?

Convex and Concave Mirrors Not all mirrors are flat, or plane. Some have curved surfaces. A mirror with a surface that curves outward is called a convex mirror. The images reflected in a convex mirror appear to be smaller and farther away. These images are always virtual. Convex mirrors are used in car side-view mirrors and for security in stores.

A mirror with a surface that curves inward is called a concave mirror. Images can be real or virtual, depending on the position of the object. Sometimes images reflected in a concave mirror will appear magnified. Concave mirrors are used in makeup mirrors and as headlight reflectors.

3 **APPLY:** Why do you think a car's side-view mirror would be convex?

CHECKING CONCEPTS

1. On a windy day, the surface of a pond will form a __________ reflection.
2. A plane mirror forms __________ images.
3. The angle of incidence is equal to the angle of __________.
4. If an incident ray is at a 45° angle, the __________ ray will also be at a 45° angle.
5. A __________ is a smooth surface that reflects light and forms images.
6. A __________ mirror has a surface that curves inward.

THINKING CRITICALLY

7. **INFER:** How does reflection help us see nonluminous objects?
8. **PREDICT:** A bathroom mirror is covered with steam. Will the mirror produce a diffuse reflection or a regular reflection? Explain.

BUILDING MATH SKILLS

Measuring the Difference Look at Figure 4-24. Measure the distance from the object to the mirror. Compare this distance with the distance from the image to the mirror. Now, compare the height of the object and the height of the image. What does this tell you about the image formed by a flat mirror?

Hands-On Activity

HOW TO MAKE A PERISCOPE

You will need scissors, heavy oak tag, rubber cement, and two round pocket mirrors.

1. Draw solid lines on your oak tag like the lines in Figure 4-25.
2. Cut along the solid lines to make the frame and slits for the mirrors. ⚠ **CAUTION:** Be careful when using scissors.
3. Fold the oak tag by following the dotted lines in the pattern. Glue or tape the long edges together.
4. Place mirrors at opposite ends of this box. One mirror should be angled toward the middle of the box; the other should be angled toward its nearest end opening. Glue the mirrors in place. Then, glue the top and bottom.
5. Now, test your periscope.

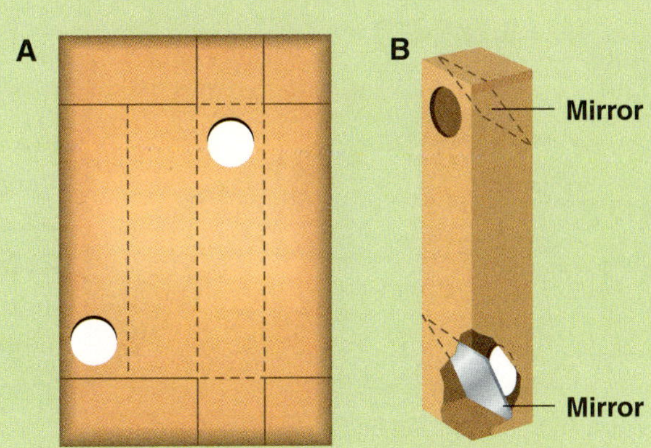

▲ Figure 4-25

Practicing Your Skills

6. **IDENTIFY:** What parts in the periscope help you see around corners?
7. **EXPLAIN:** How is this possible?

4-9 What is color?

Objectives

Describe how a prism forms a visible spectrum.

Explain why different objects have different colors.

Key Terms

prism (PRIZ-uhm)**:** triangular piece of transparent glass that separates white light into a band of colors

dispersion (di-SPUR-shun): separation of white light into its component colors

Visible Spectrum Visible light, or white light, is made up of seven different colors. These colors, which combine to make up the visible spectrum, are red, orange, yellow, green, blue, indigo, and violet. Each of these colors of light has its own wavelength. Red light has the longest wavelength. Violet light has the shortest wavelength.

DISPERSION OF WHITE LIGHT

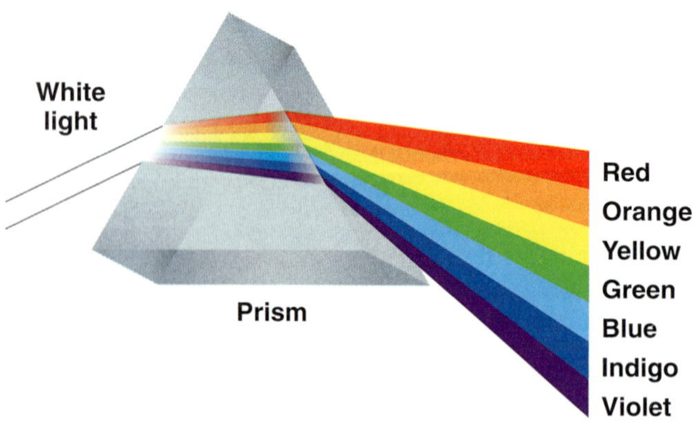

▲ **Figure 4-26** A prism separates white light into the colors of the visible spectrum.

A triangular piece of clear glass called a **prism** can be used to separate white light into its colored spectrum. Light is bent, or refracted, as it enters the prism and again as it leaves the prism. Each color of light is refracted by a slightly different amount, causing the white light to break up into its individual colors. This process of separating white light into its individual colors is called **dispersion.**

1 DESCRIBE: How does wavelength affect the color of light?

Color of Objects When white light reflects from an object, the object may appear to have a color. The object absorbs all the colors of the spectrum except the color it reflects. When white light shines on a red object, all colors except red are absorbed. The red light is reflected, and the object appears red.

▲ **Figure 4-27** An object appears red because it absorbs all colors except red.

Some objects absorb all of the light and do not reflect any color. These objects appear black. Other objects reflect all of the light. They do not absorb any of the colors. These objects appear white. Others absorb all colors but one. The color you see is the color reflected by the object.

▲ **Figure 4-28** A black object absorbs all light. None is reflected.

▲ **Figure 4-29** A white object reflects all colors.

2 PREDICT: What color will an object appear if it reflects only green light?

CHECKING CONCEPTS

1. A blue object absorbs all the colors of the visible spectrum except ___________.
2. A ___________ separates white light into the colors of the visible spectrum.
3. Red has the ___________ wavelength of all the colors of the visible spectrum.
4. An object appears ___________ because it absorbs all the colors of the visible spectrum.

THINKING CRITICALLY

5. **INFER:** The frequency of a wave increases as its wavelength decreases. Which color of light has the highest frequency? Which has the lowest?
6. **COMPARE:** How is a rainbow like a visible spectrum?

Web InfoSearch

Colorblindness The inability to see the color red, green, or blue is called colorblindness. A colorblind person sees these colors as gray. Colorblindness is a genetic disorder. It is inherited from a person's parents. Men are colorblind more often than are women.

SEARCH: Use the Internet to find out more about colorblindness. Explain why men have this disorder more often than women do. Start your search at www.conceptsandchallenges.com. Some key search words are **colorblindness, color receptors,** and **colorblind genes.**

Integrating Life Science

TOPICS: animals, ecology, adaptations, camouflage

COLOR IN ANIMALS

A pigment is a substance found in living things that absorbs light. Different pigments absorb different wavelengths of light and reflect others.

Animals contain pigments in their skin that cause them to be a certain color. These pigments are contained in special cells called chromatophores. Many animals have colors that help them blend into their surroundings. This is called camouflage. Camouflage is an adaptation that allows the animal to hide itself from predators. For example, the flounder has coloring that closely matches the color of the sea floor.

▲ **Figure 4-30** A green tree frog is well hidden in the leaves.

Not all animals try to blend in. Sometimes organisms do just the opposite. Some animals, especially those that are poisonous, have very bright colors. These colors are called warning colors. Poison dart frogs from the South American rain forest have been found in a wide array of colors and patterns. These frogs can be found in blue, yellow, orange, and even purple. These colors warn predators, "Don't eat me. I'm poisonous!"

Thinking Critically Some frogs of the rain forest are brightly colored but are not poisonous. What do you think is the reason for this?

4-10 How can light be used?

Objective

Identify uses of light in different areas of life.

Key Terms

laser: device that produces a powerful beam of light

total internal reflection: occurs when light is repeatedly reflected within a given material

fiber optics: use of optical fibers

Lasers Have you ever heard of a powerful beam of light that can cut through a steel plate? This kind of light is different from white light. This light comes from a laser. A **laser** is a device that produces a very powerful beam of light.

The difference between white light and laser light is their wavelengths and energy content. White light is made up of many different wavelengths. It includes all the visible colors. Laser light is made up of only one wavelength. As a result, laser light is only one color. Unlike the light waves in white light, the waves in laser light are all in step. A beam of laser light can travel long distances in a straight line. It does not spread out as a beam of white light does.

1 DEFINE: How is laser light different from normal white light?

Uses of Lasers Lasers have many uses. In medicine, lasers can be used to repair a detached retina and prevent blindness. A detached retina results when the retina comes loose from the back of the eye. The heat of a laser beam can attach the retina back into place. Lasers are also used to treat skin tumors, birthmarks, and some cancers.

▲ **Figure 4-31** A laser beam cutting metal

In industry, lasers can be used for cutting, welding, and drilling. For example, a laser beam can make a clean cut through thick layers of cloth. You are probably familiar with the use of lasers in CDs and DVDs. Lasers are also used in supermarkets to read the bar codes on many different products.

2 EXPLAIN: How can a laser repair a detached retina?

Optical Fibers Glasslike fibers that can transmit information in the form of light pulses are called optical fibers. These fibers are able to send information much more quickly and clearly than copper wire or radio waves can.

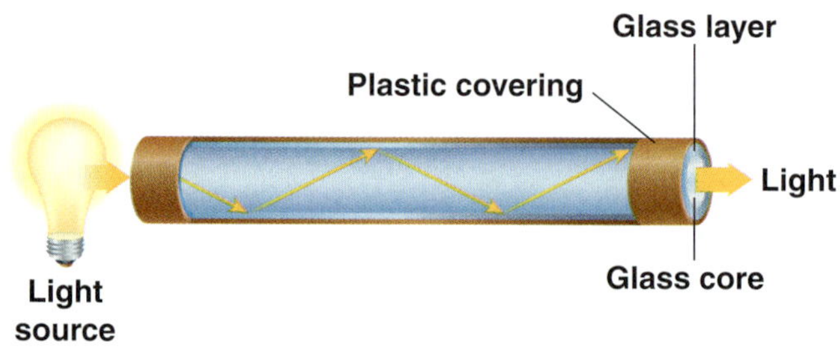

▲ **Figure 4-32** Total internal reflection

The reason light travels through optical fibers so quickly is because it reflects within the glass. If light is reflected at a great enough angle, the light does not refract through the glass. Instead, it reflects repeatedly within the optical fiber. This is called **total internal reflection.**

3 DESCRIBE: How does light travel so quickly through an optical fiber?

Uses of Fiber Optics The use of optical fibers is called **fiber optics.** Fiber optics is used in many industries. One industry using fiber optics more and more is the communications industry.

Phone companies use fiber optics in most of their new wiring. They also are replacing their remaining copper wiring with fiber optics. Cable television companies have also begun to use fiber optics. Television signals are sent to entire communities using optical fibers. Then, the signal is split into regular cable wiring, which is sent to each home.

The computer industry is also using fiber optics. Many universities and businesses use fiber optics in their local area networks. A local area network, or LAN, is a group of connected computers that share data, software, and hardware.

Fiber optics is used in medicine. Doctors can attach cameras to optical fibers. The optical fibers are then inserted into the body through a small incision, such as in the knee. This allows doctors to see within the body so they can operate making small incisions. This allows patients to recover more quickly from surgery.

4 **EXPLAIN:** How do fiber optics help patients and doctors?

CHECKING CONCEPTS

1. What is a laser?
2. What can lasers be used for in industry?
3. In what form do optical fibers transmit information?
4. What is fiber optics?

THINKING CRITICALLY

5. **CONCLUDE:** How does total internal reflection help make optical fibers work so well?
6. **INFER:** Why can the light from a laser travel a greater distance than light from a normal source of white light?

BUILDING SCIENCE SKILLS

Inferring One major difference between optical fibers and copper wiring is the way that they send and receive information. Optical fibers transmit information using pulses of light. Copper wiring uses pulses of electricity. This difference makes optical fibers less sensitive to power surges and electrical interference. Why do you think this makes optical fibers more useful in computers and computer networks?

Science and Technology

HOLOGRAMS

There is a special kind of photography that uses laser light. This process is known as holography (huh-LAHG-ruh-fee). Laser light is used to produce a three-dimensional image. The image is known as a hologram (HAHL-uh-gram).

With the invention of the laser in 1960, Dennis Gabor, the hologram inventor, was finally able to produce crisp holographic images. He was awarded the Nobel Prize in 1971 for his work.

▲ **Figure 4-33** Holograms use laser light to create a three-dimensional effect.

To make a hologram, two pictures of an object are taken. Each picture is taken from a slightly different angle. Then, the two pictures are combined into one hologram.

When you look at a hologram, you can view the original object from different angles. It is as if the object were in front of you, even though you are seeing only an image of the object.

There are several practical uses for holograms. Architects can use holograms to show three-dimensional views of buildings. Structural engineers can use holograms to test the sturdiness of a solid object.

Thinking Critically How did the invention of the laser help Gabor complete his design of the hologram?

Integrating Technology

THE Big IDEA

How is light used in the home?

Technology is the practical application of knowledge. Beginning with the first stone tools, people have used technology to accomplish tasks. A new technology often improves on older methods. It may be faster, save energy, and produce better results. Society must always weigh the risks of new technologies against their benefits. They must always be environmentally friendly.

Advances in technology have expanded our choices at home. We can chop with a knife or a food processor. We can cook with fire, electricity, or microwaves. Newer "flash bake" ovens use light to cook foods rapidly.

We also have more forms of entertainment and communication. Music on CDs sounds clearer than music on records or tapes. Besides broadcast television, choices today include cable, satellites, videotapes, and digital versatile discs (DVDs). Fiber-optic cable makes long-distance phone calls clearer. The Internet has created another way to communicate with friends and family far away.

Like lamps and mirrors, many types of technology at home use light. Look at the boxes of text that appear on this page and the next. They point out devices that are typical in homes today and some that may become common in the future. Then, follow the directions in the Science Log to learn more about "the big idea."✦

▲ **Figure 4-34** Inside a house

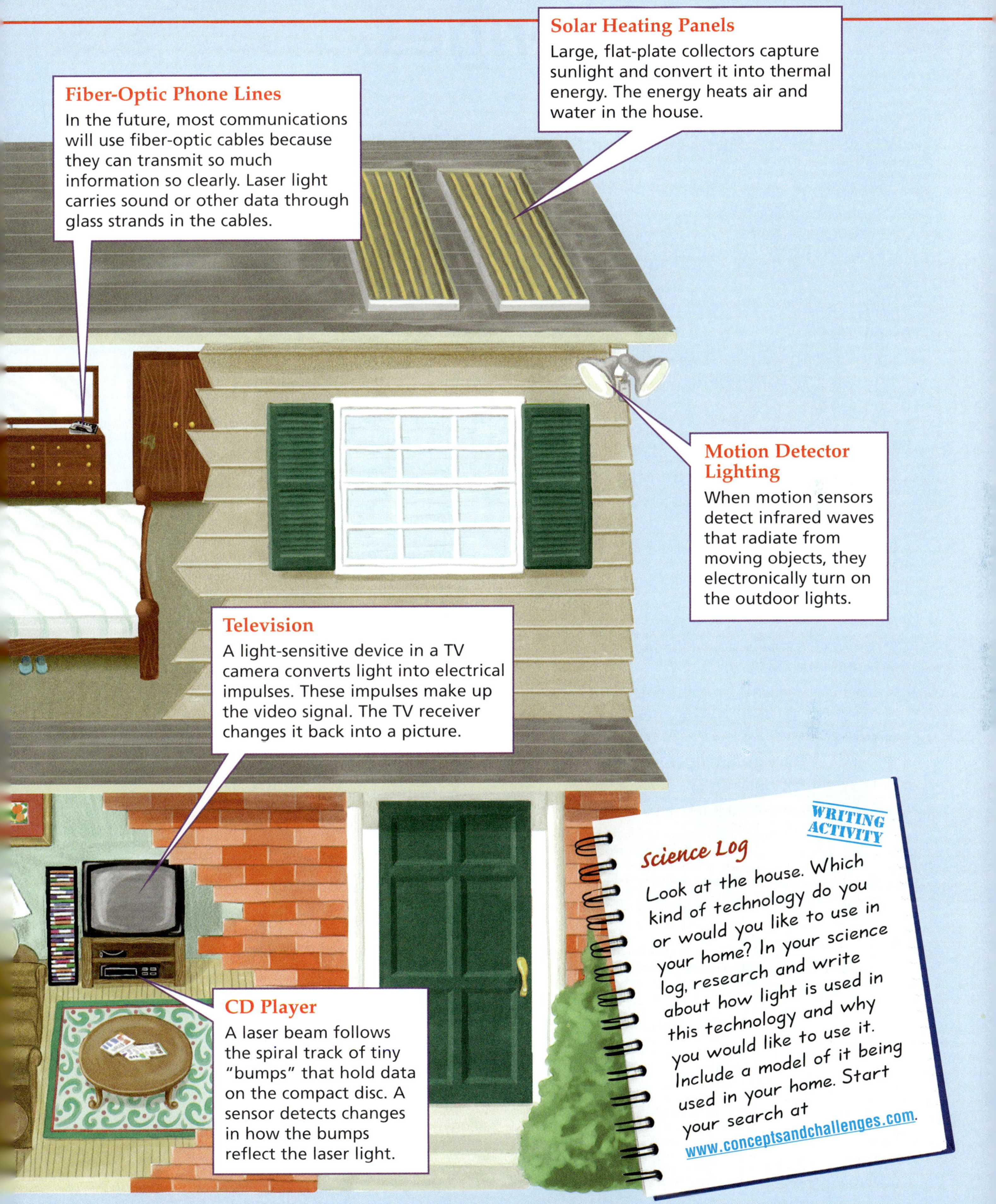

Solar Heating Panels

Large, flat-plate collectors capture sunlight and convert it into thermal energy. The energy heats air and water in the house.

Fiber-Optic Phone Lines

In the future, most communications will use fiber-optic cables because they can transmit so much information so clearly. Laser light carries sound or other data through glass strands in the cables.

Motion Detector Lighting

When motion sensors detect infrared waves that radiate from moving objects, they electronically turn on the outdoor lights.

Television

A light-sensitive device in a TV camera converts light into electrical impulses. These impulses make up the video signal. The TV receiver changes it back into a picture.

CD Player

A laser beam follows the spiral track of tiny "bumps" that hold data on the compact disc. A sensor detects changes in how the bumps reflect the laser light.

WRITING ACTIVITY

Science Log

Look at the house. Which kind of technology do you or would you like to use in your home? In your science log, research and write about how light is used in this technology and why you would like to use it. Include a model of it being used in your home. Start your search at www.conceptsandchallenges.com.

Chapter 4 Challenges

Chapter Summary

Lesson 4-1

- The **electromagnetic spectrum** includes radio waves, microwaves, infrared waves, visible light, ultraviolet rays, X-rays, and gamma rays.

Lesson 4-2

- **Light** is a form of electromagnetic energy.

Lesson 4-3

- Light travels in **transverse waves**. Light does not need a medium in which to travel.

Lesson 4-4

- **Luminous objects** give off their own light. **Illuminated objects** reflect light.
- **Opaque, transparent,** and **translucent** are three types of objects that affect light differently.

Lesson 4-5

- **Chlorophyll** is a substance in plants that absorbs the Sun's light and allows **photosynthesis** to occur.

Lesson 4-6

- A **convex lens** curves outward, whereas a **concave lens** curves inward.
- A convex lens is used to form **real images.** A concave lens can form **virtual images.**

Lesson 4-7

- The parts of the **eye** are the cornea, pupil, iris, lens, retina, and optic nerve.
- Light is focused onto the retina by the lens. The brain interprets the **image** on the retina to make it look right-side up.

Lesson 4-8

- A **plane mirror** is a smooth surface that reflects light and forms images.
- **Regular reflection** produces a clear image. **Diffuse reflection** produces a fuzzy image.

Lesson 4-9

- A **prism** can separate white light into the colors of the spectrum.
- The color of an object depends on what part of the spectrum it reflects or absorbs.

Lesson 4-10

- A **laser** produces a powerful beam of light.
- Optical fibers are glass fibers that can quickly transmit information in the form of light pulses.

Key Term Challenges

chlorophyll (p. 92)
chloroplast (p. 92)
concave lens (p. 94)
convex lens (p. 94)
diffuse reflection (p. 100)
electromagnetic spectrum (p. 84)
eye (p. 96)
fiber optics (p. 104)
illuminated object (p. 90)
image (p. 96)
laser (p. 104)
lens (p. 94)
light (p. 86)
luminous object (p. 90)
opaque (p. 90)
photon (p. 86)
photosynthesis (p. 92)
plane mirror (p. 100)
prism (p. 102)
ray (p. 86)
real image (p. 94)
regular reflection (p. 100)
sight (p. 96)
total internal reflection (p. 104)
translucent (p. 90)
transparent (p. 90)
transverse wave (p. 88)
virtual image (p. 94)
visible spectrum (p. 84)

MATCHING Write the Key Term from above that best matches each description.

1. organ used for sight
2. straight line that shows the direction of light travel
3. image that can be projected
4. seven colors that make up white light
5. small bundle of energy
6. material that allows light to pass through easily
7. device that produces a powerful beam of light
8. range of electromagnetic waves
9. reflection that forms a fuzzy image

FILL IN Write the Key Term from above that best completes each statement.

10. By the process of____________, a plant makes its own food from the Sun's light.
11. When you shine a light in a dark room, the objects that you see are ____________.
12. An ____________ forms on the retina.
13. Lamps are ____________ because they give off their own light.
14. A ____________ is a smooth surface that reflects and refracts images.

Content Challenges TEST PREP

MULTIPLE CHOICE **Write the letter of the term or phrase that best completes each statement.**

1. Chlorophyll absorbs all the colors of the visible spectrum except
   a. red.
   b. green.
   c. blue.
   d. violet.

2. The refracted image from a convex lens is
   a. right-side up.
   b. reversed.
   c. upside down.
   d. none of the above.

3. The color that is not one of the seven colors of the visible spectrum is
   a. red.
   b. blue.
   c. indigo.
   d. brown.

4. The type of electromagnetic wave that can be felt as heat is
   a. gamma rays.
   b. infrared rays.
   c. X-rays.
   d. ultraviolet rays.

5. The part of the eye that controls the amount of light entering the eye is the
   a. pupil.
   b. cornea.
   c. iris.
   d. retina.

6. In photosynthesis, electromagnetic energy is converted to
   a. electrical energy.
   b. nuclear energy.
   c. chemical energy.
   d. mechanical energy.

7. Unlike a beam of white light, a laser beam has
   a. no color.
   b. one wavelength.
   c. two colors.
   d. no wavelength.

8. The point where the light that passes through a lens is brought together is called the
   a. focal length.
   b. refraction point.
   c. focal point.
   d. retina.

9. Light reflects completely within a fiber-optic cable because of
   a. gravity.
   b. total internal reflection.
   c. photosynthesis.
   d. regular reflection.

10. An object that absorbs all colors of light and does not reflect any appears
   a. green.
   b. white.
   c. black.
   d. blue.

TRUE/FALSE **Write *true* if the statement is true. If the statement is false, change the underlined term to make the statement true.**

11. According to the law of refraction, the angle of incidence is equal to the angle of reflection.
12. Illuminated objects, such as the Sun, reflect light.
13. The reason light travels through optical fibers so quickly is because it refracts within the glass.
14. The parts of a cell where photosynthesis takes place are called chloroplasts.
15. All electromagnetic waves are streams of photons.

Concept Challenges TEST PREP

WRITTEN RESPONSE **Answer each of the following questions in complete sentences.**

1. **COMPARE:** What are two ways that a photon is similar to a particle?
2. **EXPLAIN:** How does a prism separate white light into a spectrum?
3. **INFER:** Why do electromagnetic waves travel best in a vacuum?
4. **EXPLAIN:** Why does a laser beam not spread out over long distances?
5. **HYPOTHESIZE:** Some animals can hear ultrasonic sounds, even though humans cannot. Do you think that some animals can see ultraviolet light, even though humans cannot? Why or why not?
6. **MODEL:** Think of your eye as a camera. A camera has an opening for light to enter called an aperture. Which part of the eye is similar to the camera's aperture? Which part is similar to the film in the camera? Which part is similar to the camera lens?

INTERPRETING A DIAGRAM **Use Figure 4-35 to answer the following questions and to complete the exercise.**

7. What type of lens is shown?
8. Which letter on the diagram represents the focal length of the lens?
9. Which letter represents the focal point?
10. Copy the diagram on a sheet of paper. Show the path of light rays passing through the lens.
11. Are the light rays brought together or spread apart by the lens?
12. What kind of image would be formed at point C?

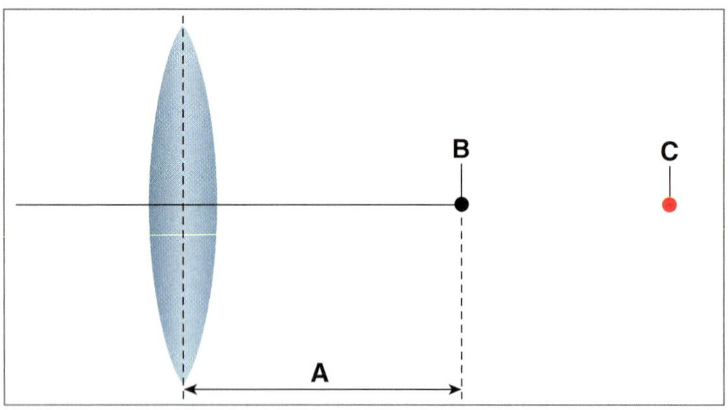

▲ Figure 4-35

Chapter 5 Electricity

▲ **Figure 5-1** The skyline at night in Seattle, Washington

Electricity provides the energy that lights up the skyline of the city in Figure 5-1. Electricity can make a city look beautiful at night, but it also plays a very important role in your life. From providing power to the appliances in your home to helping your body function properly, you rely on electricity. Although it is a very useful source of energy, electricity can be dangerous. Safety rules must be followed when using electricity.

▶Can you name some safety rules to follow when using electricity?

Contents

5-1 What is electric charge?

INVESTIGATE
Observing Electric Charges
HANDS-ON ACTIVITY

1. Blow up a rubber balloon. Tie the end into a knot.
2. Rub the balloon with a piece of wool or nylon.
3. Dim the lights in your classroom. Bring your finger near the balloon. Observe what happens between the balloon and your finger.

THINK ABOUT IT: What did you observe between your finger and the balloon? Has something like this ever happened to you before?

Objective
Explain how objects become electrically charged.

Key Terms
proton: atomic particle with a positive electric charge

electron: atomic particle with a negative electric charge

neutron: atomic particle with neither a negative nor a positive electric charge

static electricity: buildup of electric charges in an object

Atoms You have learned that all matter is made up of atoms. The three basic parts of an atom are protons, neutrons, and electrons. Protons and neutrons are found in the nucleus of an atom, and electrons are found in the electron cloud that surrounds the nucleus.

1 NAME: What are the three main parts of an atom?

Electric Charge Protons and electrons have a property called electric charge. **Protons** have positive electric charge (+). **Electrons** have negative electric charge (–). The strength of the positive electric charge on a proton is the same as the strength of the negative electric charge on an electron. **Neutrons** do not have charge. They are neutral.

Electric charge is not a physical property that can be seen or touched. However, electric charge exerts a force that affects the behavior of matter. Matter that is electrically charged follows these rules:

- Objects with like charges repel, or push away, from each other.
- Objects with unlike charges attract each other.

Because they have opposite charges, protons and electrons attract each other. However, because they have the same electric charge, protons do not attract each other. They repel, or push away, from each other. Electrons also repel each other.

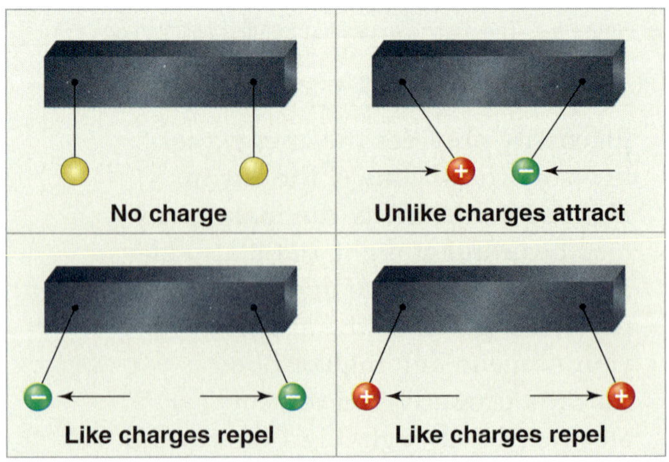

▲ **Figure 5-2** How electric charges behave

2 EXPLAIN: How do protons and electrons affect each other?

Neutral Objects Although an atom is made up of electrically charged particles, the atom as a whole has no electric charge. The number of protons and the number of electrons in an atom is always the same. Therefore, the electric charges cancel each other. The atom is neutral.

Most matter is electrically neutral. However, neutral objects can become electrically charged. An object becomes electrically charged when its atoms lose or gain electrons. The electrons in the atoms of some matter are held loosely to the atoms. If these electrons are separated from their atoms, the atoms become electrically charged. For example, when you walk across a carpet, your feet can rub off and pick up electrons from the carpet. Your body then gets a small electric charge. If you then touch a metal doorknob, electrons can jump from your hand to the doorknob. You can see a spark.

3 **DESCRIBE:** Why are atoms usually neutral?

Static Electricity Protons are held tightly in the nucleus of all atoms. However, electrons are free to move from a region of negative electric charge to a region of positive electric charge. The buildup of electric charges on an object is called **static electricity**. The word *static* means "not moving." In static electricity, electrons do not flow. They remain at rest for a time but will eventually leave the object.

Static electricity can be produced when different materials rub against each other. For example, if a sock and a shirt rub against each other in a clothes dryer, electrons may be transferred between both items. The sock may lose some of its electrons to the shirt and become positively charged. The shirt gains these electrons and becomes negatively charged. Because the sock and the shirt have opposite charges, they stick to each other. This attraction is called static cling.

4 **INFER:** Why can a sock stick to a shirt after coming out of a clothes dryer?

Static Discharge Objects that have static electricity will eventually lose their charge. The extra electrons will move onto other objects, and the original object will return to its neutral condition. The loss of static electricity is called static discharge.

A dramatic example of static discharge is lightning. During a storm, water particles in clouds swirl around and rub against each other. This friction produces electric charges in different areas of the clouds. The negative area of the cloud repels the negatively charged electrons on the surface of the ground below. The surface of the ground then becomes positively charged. An attraction develops between the negatively charged area of the cloud and the positively charged ground. If the force of attraction becomes great enough, electrons from the cloud will jump to the ground. When this happens, a spark of lightning is produced.

▲ **Figure 5-3** The force of attraction between charges in clouds and the ground can produce lightning.

5 **EXPLAIN:** How does lightning occur?

CHECKING CONCEPTS

1. An electron has a __________ charge.
2. A __________ object has neither a positive nor a negative charge.
3. Static electricity is the buildup of __________ on an object.
4. Two electrons will __________ each other.

THINKING CRITICALLY

5. **EXPLAIN:** What can happen if you walk across a carpet and then touch a metal doorknob?
6. **INFER:** Why is the buildup of electric charge on an object called static electricity?

INTERPRETING VISUALS

Use Figure 5-2 to answer the following questions.

7. **INFER:** How do like charges behave?
8. **EXPLAIN:** How do objects without a charge behave?

LAB ACTIVITY
Observing Static Electricity

BACKGROUND

Static electricity is produced when different materials rub against each other. Electrons from one object may be gained by another object. The object that loses electrons becomes positively charged. The object that gains electrons becomes negatively charged. The opposite charges cause the two objects to stick to each other.

PURPOSE

In this activity, you will observe how static electricity can attract an object.

PROCEDURE

1. Copy the chart in Figure 5-4 onto a sheet of paper.
2. Use a compass to lightly draw a large circle on a sheet of tracing paper. The circle should be about 20 cm in diameter. Be careful not to tear the tracing paper.
3. Use scissors to cut the circle out of the tracing paper. Place the tip of a pen in the center of the circle. Draw a spiral that goes all the way to the very edge of the circle. Again, draw carefully so that you do not tear the tracing paper.
4. Carefully cut along the spiral line that you drew. When you are finished, you should have one long continuous piece of tracing paper. Lay the spiral in a metal pie plate.
5. Place the end of your pen above the center of the spiral without touching the tracing paper. Observe what happens. Record your observations on the chart.
6. Rub your pen with a piece of silk a few times. Again, place the end of the pen slightly above the center of the spiral. Do not touch the tracing paper. Slowly lift your pen up. Observe what happens. Record your observations.

▲ **STEP 2** Draw a 20-cm-wide circle on tracing paper.

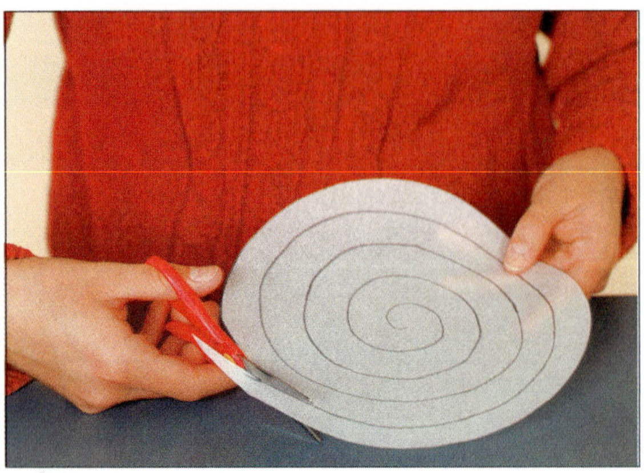

▲ **STEP 4** Cut along the spiral line.

▲ **STEP 5** Place your pen above the center of the spiral.

▲ **STEP 6** Rub your pen with a piece of silk.

Observing Static Electricity

Experiment	Observations
Pen above spiral	
Pen above spiral after rubbing with a piece of silk	

▲ **Figure 5-4** Copy this chart and use it to record your observations.

CONCLUSIONS

1. **OBSERVE:** What happened the first time you placed your pen above the spiral tracing paper? What happened after you rubbed the pen with the piece of silk?
2. **ANALYZE:** Why was there a different result?
3. **INFER:** What was the purpose of rubbing the pen with the piece of silk?
4. **INFER:** Why did you use tracing paper instead of regular paper?
5. **HYPOTHESIZE:** After Step 6, what eventually happened to the spiral tracing paper?

5-2 What is a battery?

INVESTIGATE

Making a Lemon Wet Cell
HANDS-ON ACTIVITY

1. Wrap several turns of wire around a magnetic compass.
2. Wrap one end of the wire around the end of a strip of copper. Wrap the other end of the wire around the end of a strip of zinc.
3. Insert the free end of the copper strip into the side of a lemon. Insert the free end of the zinc strip next to the copper strip. Make sure that the strips do not touch each other.

STEP 3

THINK ABOUT IT: What happens to the needle of the compass? What do you think takes place?

Objectives

Identify a battery as a series of electrochemical cells that are connected together. Compare a wet cell and a dry cell.

Key Terms

electrochemical cell: device that changes chemical energy into electrical energy

battery: a series of electrochemical cells that are connected to each other

electrolyte: substance that forms ions when melted or dissolved in water

Electrochemical Cell Under certain conditions, chemical energy can be converted to electrical energy. A device that changes chemical energy to electrical energy is called an **electrochemical cell.** A series of two or more electrochemical cells that are connected to each other is called a **battery.**

1 DEFINE: What is a battery?

Wet Cell The simplest type of electrochemical cell is called a wet cell. A wet cell is made up of a negative pole, a positive pole, and an electrolyte. An **electrolyte** is a substance that forms ions when melted or dissolved in water. The electrolyte in most wet cells is an acid or a base. The negative and positive poles of a wet cell are called electrodes. The ions transfer the negative charge between the electrodes in a cell. Zinc can be used as a negative electrode. Copper can be used as a positive electrode.

A chemical reaction in a wet cell causes electrons to move from the copper electrode to the zinc electrode. As Figure 5-5 shows, when the two electrodes are connected to each other by a metal wire, electrons flow from the zinc electrode through the wire to the copper electrode.

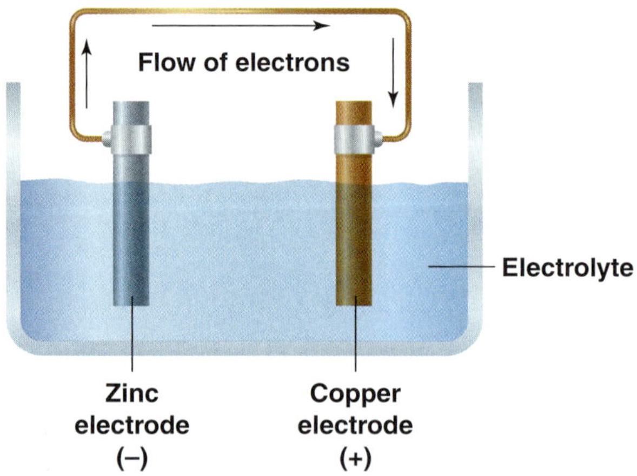

▲ **Figure 5-5** Outside a wet cell, electrons flow through a wire from the negative electrode to the positive electrode.

2 IDENTIFY: What are the three parts of a wet cell?

Dry Cell A dry cell is used as a source of electrical energy in flashlights and portable electronic devices. A dry cell works in the same way that a wet cell does. The electrolyte is a moist paste inside the cell. The zinc casing of the cell is the negative electrode. The positive electrode runs down the center of the battery and is usually made of carbon.

When the two electrodes are connected by a conductor, electrons flow through the conductor from the carbon to the zinc.

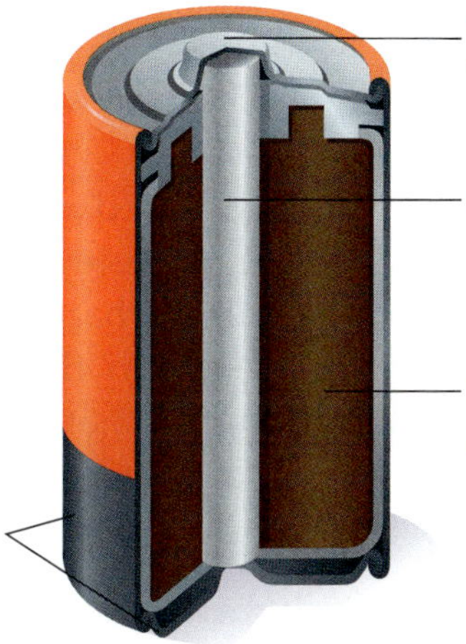

◀ **Figure 5-6** In a dry cell, the electrolyte is a moist paste.

3 **CONTRAST:** How is the electrolyte in a dry cell different from that in a wet cell?

CHECKING CONCEPTS

1. A series of electrochemical cells that are connected to each other makes up a __________.
2. A __________ cell is used in a flashlight.
3. An electrochemical cell converts chemical energy into __________ energy.
4. The negative electrode of a wet cell is usually made of __________.

THINKING CRITICALLY

5. **ANALYZE:** In a wet cell, the electrons flow from the zinc electrode to the copper electrode. Why do they not flow from the copper to the zinc?

Web InfoSearch

Rechargeable Batteries Batteries that can be used over and over again are called rechargeable batteries. Two types of rechargeable batteries are lithium batteries and nickel-cadmium batteries. These types of batteries can be used in cellular phones, laptop computers, and digital cameras.

SEARCH: Use the Internet to find out how they work. Start your search at www.conceptsandchallenges.com. Some key search words are **batteries** and **rechargeable.**

Real-Life Science

DISPOSAL OF BATTERIES

Every year, Americans use billions of batteries. When used up, most of these batteries are thrown into the trash. These batteries are then buried in landfills or they are burned in incinerators. However, batteries often contain poisonous metals that can leak into the groundwater through landfills. They can also be released into the air when batteries are incinerated, or burned. Two of these metals, mercury and cadmium, are very dangerous. Mercury can cause harm to the nervous systems in humans and in fish. Cadmium can cause cancer.

Batteries should be separated from ordinary trash and disposed of in special hazardous-waste landfills. Such landfills are built on top of clay to stop poisons from leaking into groundwater. Many cities now have special sites for the collection of batteries. To prevent the release of dangerous materials into the environment, we must dispose of batteries safely.

▲ **Figure 5-7** The proper disposal of batteries is an important environmental concern.

Thinking Critically Why is the proper disposal of batteries so important?

5-3 What are insulators and conductors?

Objective

List some examples of conductors and insulators.

Key Terms

conductor: material through which electric charge can flow easily

insulator: material through which electric charge does not flow easily

Conductors A material through which electric charge can flow easily is called a **conductor.** Most metals are good conductors of electricity.

If a metal wire is placed between the positive and negative poles of a battery, a path is created so that electrons can flow. Suppose metal wires are used to attach a light source to the positive and negative poles of a battery. Electrons will flow through the wire and the light bulb will light up. If a piece of rubber is used instead of metal wires, electrons will not flow. The light bulb will not light up.

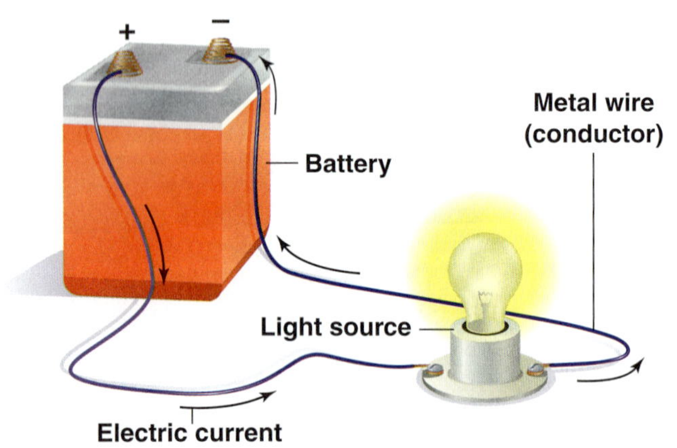

▲ **Figure 5-8** Electrons will flow if a metal wire is attached to both poles of a battery.

Some conductors are better than others. Wire made of copper is one of the best conductors of electricity. Wire made of aluminum is lighter and is less expensive than wire made of copper. However, aluminum is not as good a conductor of electricity.

1 DEFINE: What is a conductor?

Insulators A material through which electric charge does not flow easily is called an **insulator.** Rubber is an insulator. When a piece of rubber is placed between two oppositely charged objects, charges will not flow. Electrons in the atoms of rubber are held tightly together so that they cannot move freely throughout the material. Other insulators are cork, wood, cloth, plastic, and air.

2 INFER: What happens when a piece of rubber is placed between two oppositely charged objects?

Insulated Wires Electrical cords use both insulators and conductors. The conductor is usually made of copper. The copper wire is covered with an insulator, such as rubber. The rubber prevents electric charges from leaving the bare wire.

The rubber insulation of electric cords may become cracked or worn. Such electric cords are dangerous because they can cause a short circuit. A short circuit results when two uninsulated wires touch, allowing a large amount of electric charge to jump between them. Wires carrying too much charge can become very hot, perhaps causing a fire.

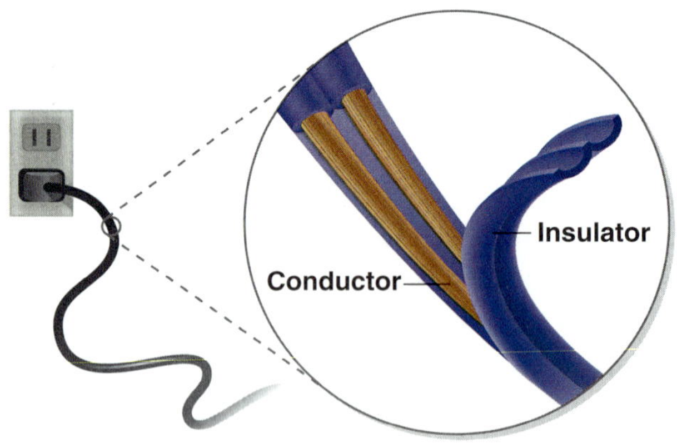

▲ **Figure 5-9** Most household cords have two wires covered with rubber insulation to prevent electric charges from leaving the wire.

3 EXPLAIN: Why are electric cords often covered with rubber?

CHECKING CONCEPTS

1. A __________ allows current to flow easily through it.
2. Rubber is an __________ because it does not allow an electric charge to flow easily through it.
3. In an insulated electrical wire, the conductor is often __________.
4. An electric cord is safe to touch because it is covered with __________.
5. If two conducting wires in an electric cord touch, a __________ will result.
6. Cork is a good __________.
7. A copper wire is one of the best __________.
8. A short circuit results when two __________ wires touch, allowing an electric charge to jump between them.

THINKING CRITICALLY

9. **INFER:** A lightning rod is supposed to keep a bolt of lightning from damaging a house. The rod is placed at the top of the house and is connected to the ground. What material would make a good lightning rod? Explain.
10. **COMPARE:** How is a conductor different from an insulator? Give an example of each.
11. **APPLY:** Why is it safe to use an insulated wire but not safe to use an uninsulated wire?

HEALTH AND SAFETY TIP

You should not try to repair or rewire electric appliances yourself. They use large amounts of electricity and can be quite dangerous. Frayed or worn electric wires can cause a bad electric shock. What should you do if wires or appliances in your home need repair?

People in Science

BENJAMIN FRANKLIN (1706–1790)

Benjamin Franklin was a politician, a writer, a businessman, and an inventor. He is probably most famous for an experiment that he performed in the summer of 1752. One night, Benjamin Franklin tied a key to the string of a kite. He then flew the kite during an electrical storm. He was hoping to collect an electric charge from lightning directly onto the key. He wanted to prove that lightning is really an electric charge sent through the air. During the experiment, Franklin was successful in conducting the charge from lightning to the key.

▲ **Figure 5-10** Benjamin Franklin conducted lightning to the ground by using a kite and a key.

Franklin performed many experiments with lightning during the course of his life. Because he understood the danger of electricity, Franklin invented the lightning rod to protect ships and buildings. He realized that lightning is attracted to tall, pointed objects. So, he invented a pointed metal object that could be mounted on the roof of tall structures. When lightning occurred, lightning bolts would be attracted to the metal rod rather than to the ship or the building.

In addition to being the first person to understand that lightning is a huge electric discharge, Benjamin Franklin also gave the names *positive* and *negative* to opposite electric charges.

Thinking Critically Why is Benjamin Franklin's experiment with the kite considered dangerous?

5-4 What are two kinds of electric current?

Objective

Differentiate between direct current and alternating current.

Key Terms

electric current: flow of electric charge through a conductor

direct current: current in which electrons always flow in the same direction

alternating current: current in which electrons constantly change direction at a regular rate

Useful Electricity You have learned that the electric charges that build up on objects with static electricity do not flow. Instead, these negative charges can jump between objects. Static electricity cannot be used to run the electrical appliances in your home. In order for electricity to be useful, the electric charges must flow steadily.

When a conductor is connected to an appliance and to the opposite poles of a battery, electrons will flow steadily through the conductor. The flow of electric charge through a conductor is called an **electric current.** An electric current is like a stream of water. However, instead of a flow of water, it is a continuous flow of electrons.

1 EXPLAIN: Why can static electricity not be used to run electrical appliances?

Direct Current An electric current in which charges flow in one direction only is called **direct current (DC).** The current produced by electrochemical cells, including dry cells, is direct current. Figure 5-11 shows two dry cells in a flashlight. When the flashlight is turned on, the electric current produced by the dry cells keeps the bulb lit. Notice that the current flows in one direction.

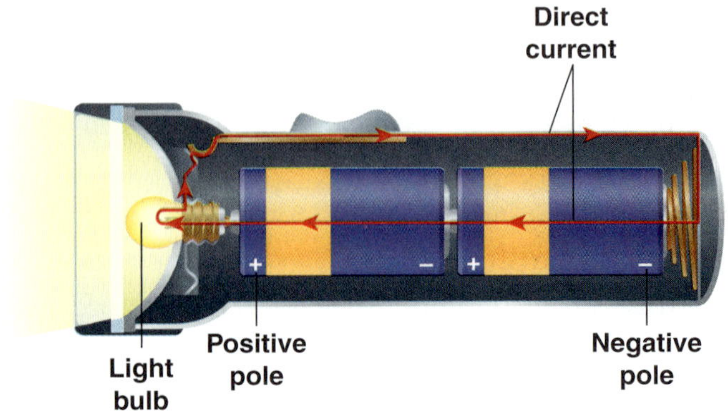

▲ **Figure 5-11** The arrows show the path followed by direct current in a flashlight.

2 DEFINE: What is direct current?

Alternating Current A second type of electric current does not flow in one direction only. An **alternating current (AC)** constantly changes direction at a regular rate. Most of the electricity found in homes, schools, and businesses is alternating current.

The main advantage of alternating current over direct current is that alternating current can be transported more easily over long distances. Another advantage is that the voltage of an alternating current can be increased or decreased.

▲ **Figure 5-12** Electricity from power plants travels as alternating current to homes.

Alternating current can be changed to direct current. In the home, many electronic devices such as computers, television sets, and CD players use direct current. Converters built into these devices are designed to change the alternating current from the household outlet to direct current.

3 DEFINE: What is alternating current?

CHECKING CONCEPTS

1. The flow of electric charge through a conductor is called __________.
2. Current that reverses direction at a regular rate is called __________ current.
3. A battery is a source of __________ current.
4. Most of the electricity that is used in everyday life comes from __________ current.
5. The type of electric current supplied by power plants is __________ current.

THINKING CRITICALLY

6. **COMPARE:** How is alternating current different from direct current?
7. **ANALYZE:** A toy manufacturer is making an electric toy that plugs into a wall outlet. The toy needs direct current in order to operate. How should the toy be made?

DESIGNING AN EXPERIMENT

Design an experiment to solve the following problem. Include a hypothesis, variables, a procedure, and a type of data to study.

PROBLEM: Will a battery last longer if it is kept in a freezer before it is used?

Integrating Life Science

TOPICS: nervous system, adaptation, predator, prey

ANIMALS THAT USE ELECTRICITY

Electric currents, or impulses, are also found in living things. The nervous system of most organisms uses electric charges. Some organisms, however, have more specialized uses of electricity.

An electric eel can stun or kill its prey with a strong electric charge. The electric eel has special musclelike cells that generate electricity. Eels also use their electric ability for self-defense against predators.

▲ **Figure 5-13** An electric eel (top) and a platypus (bottom) both use electricity to survive.

The duck-billed platypus uses electricity to find its food, even in total darkness. Duck-billed platypuses live in streams in Australia. With a special sense that detects electricity, they can find tasty crayfish and other prey, even at night in muddy water. The platypus has a special organ on the end of its soft bill that picks up faint electric signals coming from the muscles of animals swimming nearby.

Animals that produce and use electricity are usually found in water habitats. They are not found on land because water can conduct electricity, whereas air is a good insulator.

Thinking Critically Do you think that an electric eel can produce electricity out of water as well as it does in water? Explain your answer.

5-5 What is a series circuit?

INVESTIGATE

Making an Electric Circuit

HANDS-ON ACTIVITY

1. Obtain a dry-cell (size D) battery, two insulated wires, and one light bulb in a holder from your teacher.
2. Connect the light bulb holder to the battery by using the wires. You will know if you have connected everything properly when the light bulb lights up.

THINK ABOUT IT: What purpose do the wires serve? What are the three main parts to this setup?

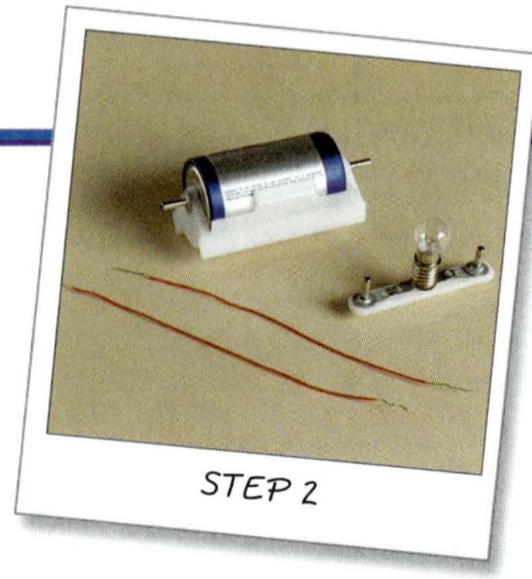

Objectives

Explain how electricity flows through a closed circuit. Describe a series circuit.

Key Terms

electric circuit: path that an electric current follows

series circuit: circuit in which electric current follows only one path

Circuits An **electric circuit** is the path that an electric current follows. All electric circuits have three parts: a source of electric energy, a load or device that uses the electric energy, and wires. The source of the electric energy can be a battery or a wall outlet. The load can be a light bulb, an appliance, or some other electric device. Wires connect the source to the load.

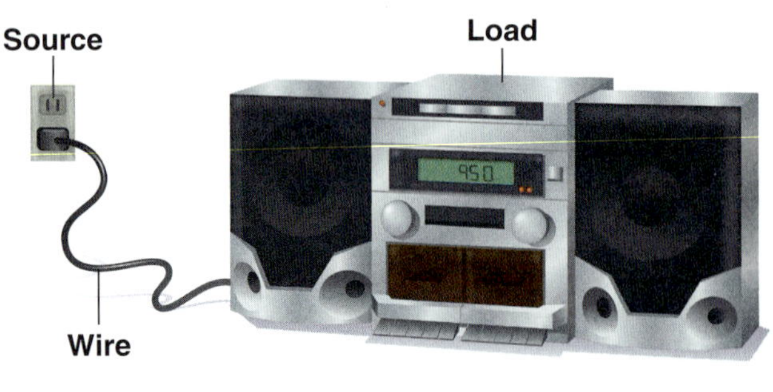

▲ **Figure 5-14** An electric circuit has three parts: a source, wires, and a load.

1 DEFINE: What is an electric circuit?

Open and Closed Circuits What would happen if the only bridge over a river was closed for repair? Vehicles using that bridge would not be able to cross the river. The path that connects both sides of the river would no longer be complete. The same situation occurs in an electric circuit. If the electric circuit is not complete, then the electric charges cannot flow. An electric circuit that is incomplete or has a break in the pathway is called an open circuit. Electric charges cannot flow through an open circuit. Electric charges can flow only through a complete or closed circuit. In a closed circuit, there are no breaks in the path.

A switch is used to control (open and close) an electric circuit. When a switch is in the "off" position, the circuit is open. Electric charges will not flow when a switch is in the off position. When the switch is in the "on" position, the circuit is closed. Electric charges will flow when the switch is in the on position.

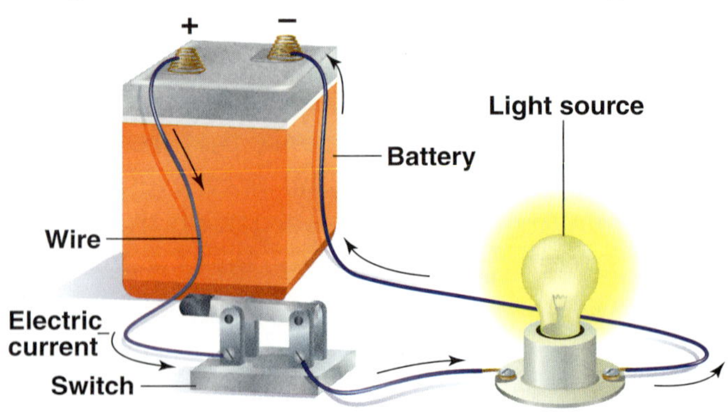

▲ **Figure 5-15** Electric charges flow when a switch is in the closed, or on, position.

2 EXPLAIN: Why is a switch used in an electric circuit?

Series Circuit The simplest type of electric circuit is called a **series circuit.** In a series circuit, the electric charges follow only one path through all elements of the circuit. Figure 5-16 shows a series circuit. In this circuit, a battery is connected to two light sources. The current goes through the first light, then through the second light, and then back to the battery.

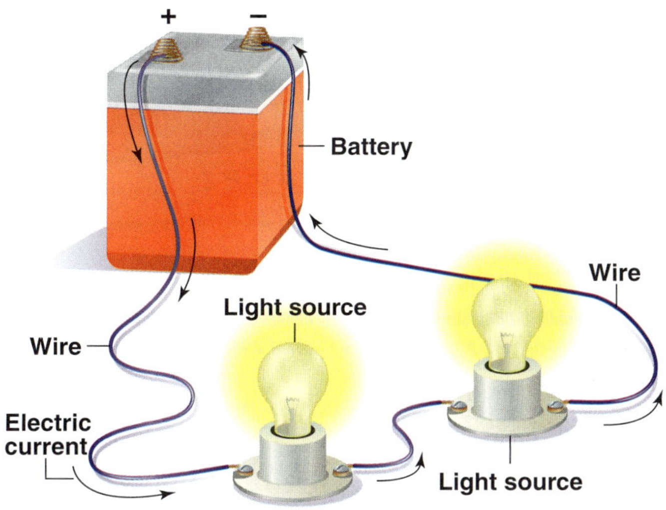

▲ **Figure 5-16** In a series circuit, electric current follows one path.

DEFINE: What is a series circuit?

CHECKING CONCEPTS

1. What is the path that an electric current follows called?
2. What are the three parts of an electric circuit?
3. In what kind of electric circuit does electric current follow only one path?
4. In which position will a switch allow electric charges to flow?

THINKING CRITICALLY

5. **HYPOTHESIZE:** In a series circuit with two lamps, what do you think would happen if one of the lamps was not working? Explain your answer.

INTERPRETING VISUALS

Use Figure 5-15 to answer the following questions.

6. **IDENTIFY:** Which object is the load?
7. **HYPOTHESIZE:** What would happen if the switch was in the off position? Explain your answer.

Science and Technology

COMPUTER CHIPS – INTEGRATED CIRCUITS

Computers are all around us. Some computers are used on a desktop, and others are used to operate microwave ovens. The main functions of a computer are to store, process, and retrieve information.

Computers process information using computer chips. Computer chips are very small pieces of a semiconductor that contain many integrated circuits. They are like miniature electric mazes with many pathways. They can perform millions of different functions. A computer's central processing unit, or CPU, may have 10 million electronic parts all packed into a 3-cm^2 wafer. Electronic signals flow through integrated circuits on a computer chip at very high speeds.

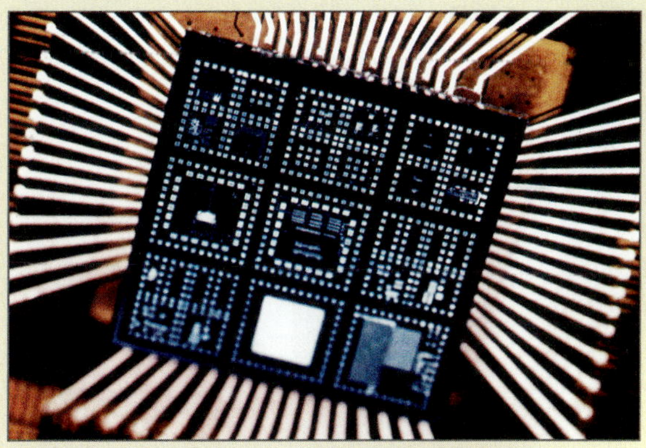

▲ **Figure 5-17** A computer chip contains many integrated circuits.

Computer chips are extremely useful, inexpensive, and small. You can find powerful chips in computers and simpler ones in everything from key chains and electronic greeting cards to cars and stereos.

Thinking Critically Why do you think computer chips are used in so many different kinds of devices?

5-6 What is a parallel circuit?

INVESTIGATE

Making Another Kind of Electric Circuit

HANDS-ON ACTIVITY

1. Obtain a dry-cell (size D) battery, four insulated wires, and two light bulbs in holders from your teacher.
2. Connect the light bulbs to the battery using the wires. Do not make a series circuit. Make sure that you use all four wires to create separate paths. You will know that you have correctly connected the wires when you remove one of the light bulbs and the other stays lit.

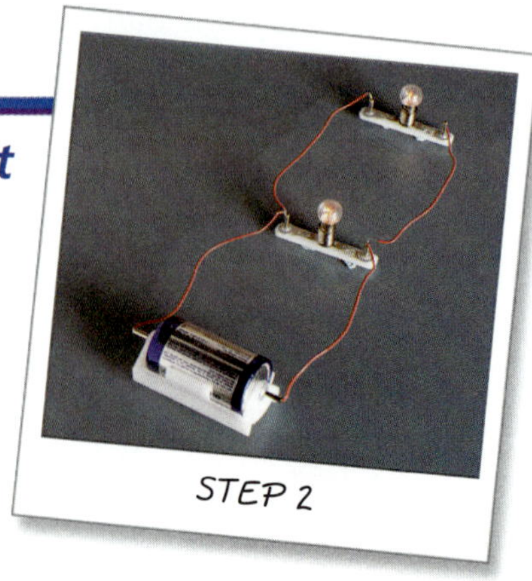
STEP 2

THINK ABOUT IT: What is the difference between this type of electric circuit and a series circuit?

Objectives

Describe a parallel circuit. Compare a parallel circuit and a series circuit.

Key Term

parallel circuit: circuit in which an electric current can follow more than one path

Parallel Circuit In a series circuit, there is only one path that an electric current can follow. In a **parallel circuit** an electric current can follow more than one path. The illustration on the right in Figure 5-18 shows a parallel circuit. Two lamps are connected to one battery. Notice how the wires are connected to each lamp. If one lamp goes out, charges can still flow through the other path. The other lamp will remain lit.

1 **DEFINE:** What is a parallel circuit?

Series Circuits versus Parallel Circuits In a series circuit with two lamps, what would happen if one of the lamps went out? The charges flowing to that lamp would stop. Because there is a break in the path, neither lamp would light. This is a great disadvantage in using a series circuit.

However, if the two lamps were set up in a parallel circuit, this problem could be avoided. In a

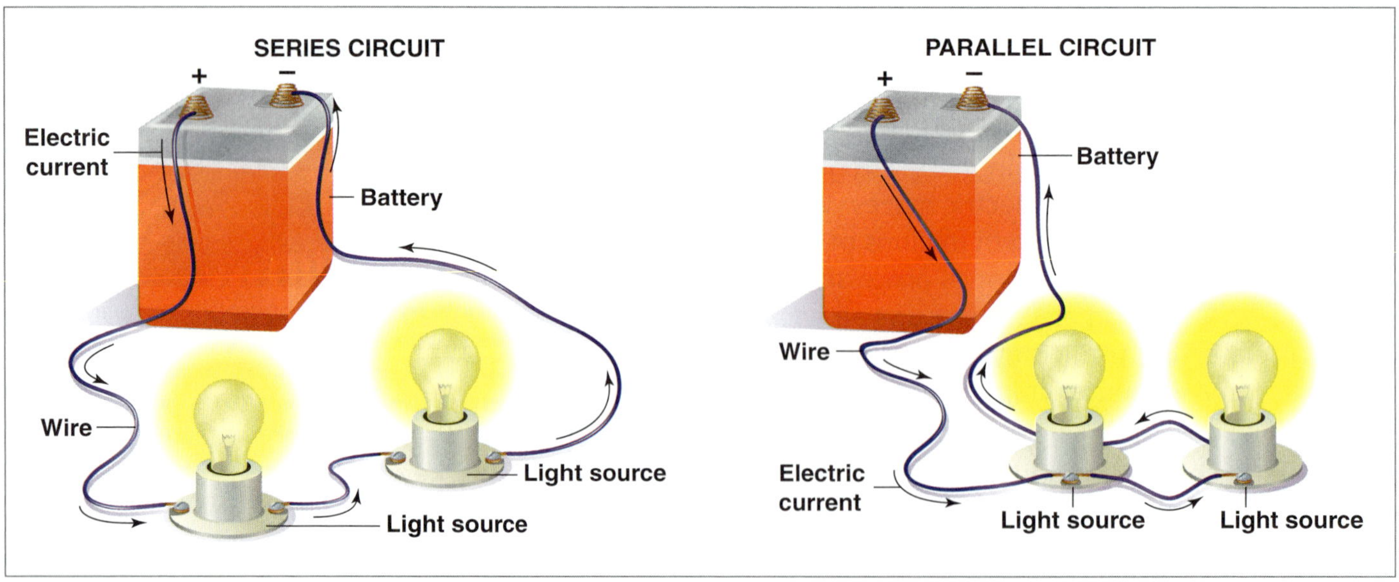

▲ **Figure 5-18** In a series circuit (left), electric current has only one path to follow. In a parallel circuit (right), electric current can follow more than one path.

parallel circuit, the current can follow more than one path. If the lamps were connected in different paths, one lamp would not stop working if the other lamp went out. Because the charges could follow different paths, electric current could still reach the other lamp. This is the reason why electric circuits in homes and businesses are parallel circuits.

2 **DESCRIBE:** What happens to the current in a series circuit when a lamp goes out?

CHECKING CONCEPTS

1. In a __________ circuit, the current follows only one path.
2. In a __________ circuit, the current can follow more than one path.
3. Homes, offices, and schools use __________ electric circuits.

THINKING CRITICALLY

4. **HYPOTHESIZE:** How could you change a series circuit into a parallel circuit?

BUILDING SCIENCE SKILLS

Labeling a Diagram Figure 5-19 shows an open series circuit with one lamp connected to the current source. Copy the diagram into your notebook. Identify and label each of the following parts in the diagram: the source of the electric energy, the lamp, the wire, and the switch.

▲ **Figure 5-19** Open series circuit

People in Science

ELECTRICIAN

An electrician is a person who is trained to install and repair electric equipment. An electrician must know how much electricity a building needs. A house with basic appliances such as a refrigerator and a television needs a certain amount of electricity. A high-rise apartment building needs much more electricity.

Most homes have wall outlets that supply a certain amount of electricity. However, different kinds of appliances may need a different amount of electricity. An electrician has to make sure that the correct amount is being used. An electrician must also be familiar with different types of wires. Most houses use copper wiring, but some use aluminum. Older homes may need to have all the wiring replaced. An electrician needs to make sure that the correct type of wiring is used.

▲ **Figure 5-20** An electrician needs to know how much electricity a building needs.

Anyone interested in a career as an electrician should take courses in mathematics, physics, and chemistry in high school. They can also participate in apprenticeship programs that take four to five years to complete.

Thinking Critically What are some of the things an electrician must know when installing wiring in a home?

5-7 How is electricity measured?

Objective

Use the correct units to measure voltage, current, and resistance.

Key Terms

voltage: energy available to move charges through a circuit

volt: unit used to measure voltage

ampere: unit used to measure electric current

resistance: opposition to the flow of electric current

ohm: unit used to measure resistance

Voltage Energy is needed to make something move, even something as tiny as an electron. The energy available to move charges through a circuit is called **voltage.** Voltage is measured in units called **volts** and is measured with an instrument called a voltmeter.

1 EXPLAIN: What is voltage?

Current The amount of electric current depends on the number of charges flowing through a wire. The unit for measuring electric current is the **ampere,** or amp. An ampere is a measure of the number of charges flowing past a point in a circuit in one second. A device used to measure the amount of electric current is called an ammeter.

2 IDENTIFY: What is an ampere a measure of?

Resistance **Resistance** is the tendency for materials to oppose the flow of electric charges. Some insulating materials, such as rubber and plastic, have very high resistance. Other materials that are good conductors, especially metals, have lower resistance. The unit for measuring resistance is the **ohm.**

Four things affect the resistance of a wire.

- **Length** The longer a wire is, the more resistance it has.
- **Diameter** The thinner a wire is, the more resistance it has.
- **Material** Wires made of poor conductors have more resistance than wires made of good conductors.
- **Temperature** As a wire gets hotter, its resistance increases.

The resistance of a material to the flow of electricity can be useful. For example, a light bulb lights up because of the resistance of the wire inside the bulb. As electric current passes through a light bulb, the wire inside the bulb resists the electric current. Its resistance to the electric current heats up the wire so that it glows.

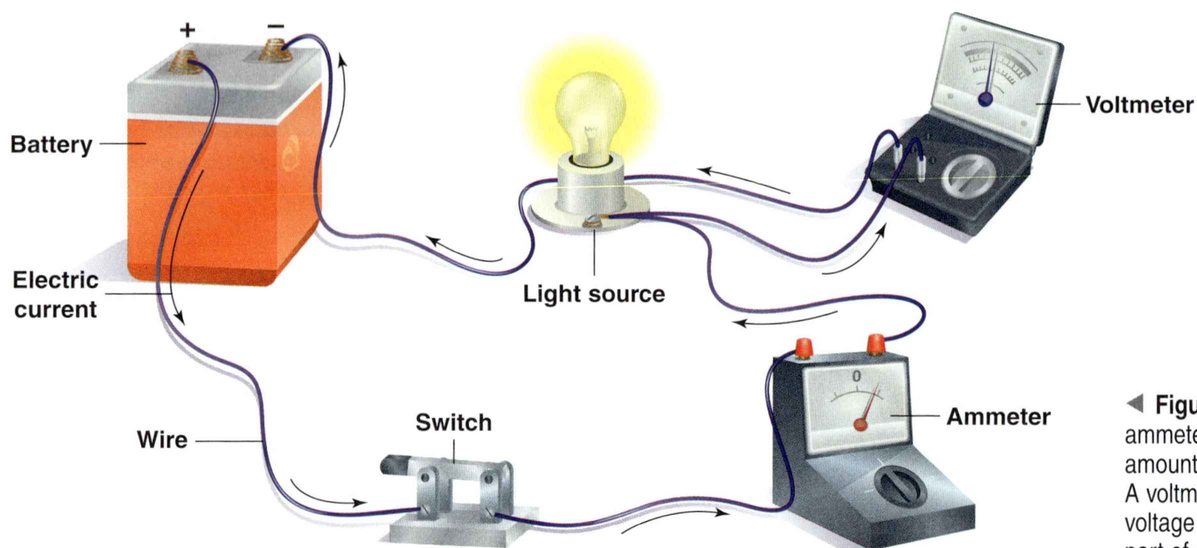

◀ **Figure 5-21** An ammeter measures the amount of electric current. A voltmeter measures the voltage drop across any part of a circuit.

◀ **Figure 5-22** Resistance to electric current causes the wire in the light bulb to glow.

3 **DEFINE:** What is resistance?

CHECKING CONCEPTS

1. Energy available to move charges through a circuit is called __________.
2. Current is measured in __________.
3. The unit for resistance is the __________.
4. A long wire has __________ resistance than a short wire.

THINKING CRITICALLY

5. **HYPOTHESIZE:** Which has more resistance, a conductor or an insulator? Explain.
6 **EXPLAIN:** How is resistance useful in a toaster?
7. **EXPLAIN:** How could you measure the amount of electric current running through a circuit?

Web InfoSearch

Volts, Amps, Ohms Volt, ampere, and ohm are units of measurements taken from the names of famous scientists.

SEARCH: Use the Internet to find out about each scientist for whom each unit of measurement is named. Start your search at www.conceptsandchallenges.com. Some key search words are **volt, ampere,** and **ohm.**

Science and Technology

SUPERCONDUCTORS

▲ **Figure 5-23** Liquid nitrogen is used to make superconductors.

When a conductor such as a wire carries an electric current, the conductor becomes hot. This causes the resistance of the conductor to increase. However, there are materials that show no resistance to the flow of electrons at temperatures close to absolute zero. Absolute zero is –273°C. Scientists are working to develop materials that have near-zero resistance at temperatures well above absolute zero. Such materials are called superconductors.

Some materials become superconductors at temperatures below –250°C. Liquid helium has been used to cool materials to this temperature. However, liquid helium is very expensive. Scientists have found that some materials become superconductors when cooled with liquid nitrogen. So, researchers are looking to use liquid nitrogen as a coolant. Liquid nitrogen does not reach temperatures as low as liquid helium, but it is much less expensive. However, much research is needed before such "warm" superconductors become part of our daily lives.

Thinking Critically Why do you think finding materials that are superconductors at room temperature can be useful?

5-8 What is Ohm's law?

Objective

Explain the relationships between electric current, voltage, and resistance.

Key Term

Ohm's law: current in a wire is equal to the voltage divided by the resistance

I, V, and R Every closed circuit has an electric current (I), voltage (V), and resistance (R). Current, voltage, and resistance vary from circuit to circuit. Different power sources have different amounts of voltage. For example, a D-cell battery has 1.5 volts, while the electricity used in our homes is at 110 volts.

◀ **Figure 5-24** Different power sources have different amounts of voltage.

Resistance in a circuit depends on the kind of wires used and the kind of load through which the current moves. A wire made of copper has less resistance than one made of tin. A toaster will offer more resistance than a mixer. The amount of current in a circuit depends on the voltage and on the total resistance offered by the load.

1 NAME: What do I, V, and R stand for?

Ohm's Law Even though the current, voltage, and resistance vary from circuit to circuit, there is often a simple relationship among them. This relationship is called Ohm's law. **Ohm's law** states that the current (I) in a wire is equal to the voltage (V) divided by the resistance (R).

Current = Voltage ÷ Resistance
or
$I = V \div R$

The units of measurements used would be amperes, which is shown by the following formula.

Amperes = Volts ÷ Ohms

2 DESCRIBE: What does Ohm's law state?

Using Ohm's Law Suppose a 12-volt battery is connected to a circuit with a resistance of 6 ohms. What is the current?

$I = V \div R$
$I = 6 \text{ volts} \div 3 \text{ ohms}$
$I = 2 \text{ amps}$

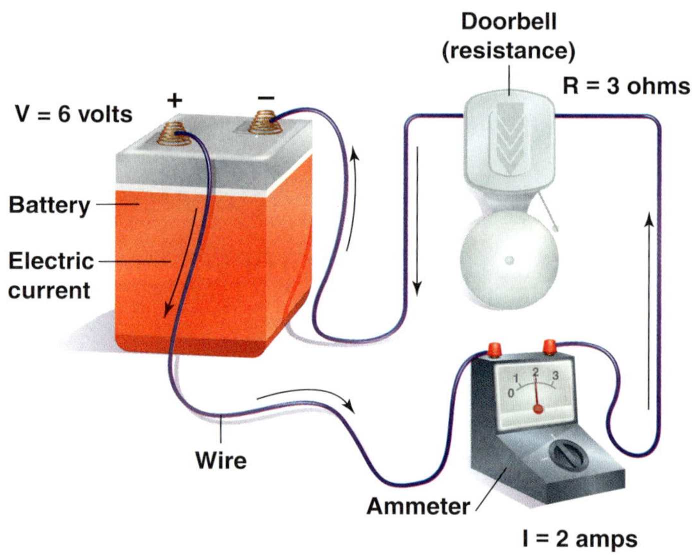

▲ **Figure 5-25** Applying Ohm's law

You can rearrange the formula so that you can find resistance or voltage. To find the voltage (V), multiply the current (I) by the resistance (R). To find the resistance, divide the voltage by the current. As long as you know any two values, you can use Ohm's law to find the remaining value.

Voltage	Resistance
$V = I \times R$	$R = V \div I$

▲ **Figure 5-26**

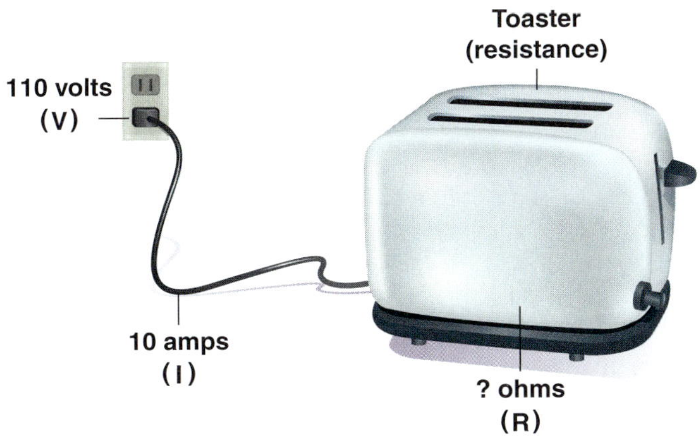

▲ **Figure 5-27** Find the resistance at the moment when I = 10 amps and V = 110 volts.

3 **CALCULATE:** What is the resistance of the toaster in Figure 5-27?

CHECKING CONCEPTS

1. Every closed circuit has an electric current, voltage, and ______.
2. According to Ohm's law, I = ______ ÷ R.
3. Resistance is measured in ______.
4. Electric current is measured in ______.

THINKING CRITICALLY

5. **INFER:** How is electric current related to resistance and voltage?

BUILDING MATH SKILLS

Calculating Copy the table in Figure 5-28 on a sheet of paper. Use Ohm's law (I = V ÷ R) to complete the chart. Express your answer in whole numbers or decimals.

I (amps)	V (volts)	R (ohms)
?	10	70
15	250	?
5	?	25
5	110	?

▲ **Figure 5-28** Find the missing values.

Science and Technology

ELECTRIC CARS

Today's gasoline-powered cars use up much of America's oil supply. The gases that they give off account for almost half the amount of pollution in the air. However, the supply of oil and other resources are limited, and the increasing pollution in the air is a health concern. As a result, some automakers are looking to make electric cars an alternative to gasoline-powered cars.

▲ **Figure 5-29** An electric car

Electric cars would reduce the use of gasoline and oil. Because electric cars do not rely on the burning of fuels, they also would create less pollution. Electric cars are also more efficient than gas-powered cars. An electric car does not waste as much of the power it produces as a gas-powered car does.

The electric cars available today use lead-acid batteries. These batteries can be charged using household current from a wall plug. However, these batteries are expensive, and they limit the distance that the car can travel. Instead of relying on batteries, some cars are being built with parts that store large amounts of electricity. Manufacturers are also looking to develop fuel cells that use hydrogen to generate electricity.

Thinking Critically Why are electric cars better for the environment than gas-powered cars?

5-9 How can you use electricity safely?

Objective

Describe ways to use electricity safely.

Key Terms

fuse: thin piece of metal that melts and breaks a circuit if too much current is flowing

circuit breaker: switch that opens a circuit if too much current is flowing

Overloaded Circuits Electricity is a wonderful source of energy, but it can also be dangerous. Too many appliances connected to one outlet may cause an overloaded circuit. In an overloaded circuit, too much current flows through a wire. The wire may overheat. It can then melt its insulation and cause a short circuit. Heat and sparks could result in a fire. A frayed wire can also cause a fire.

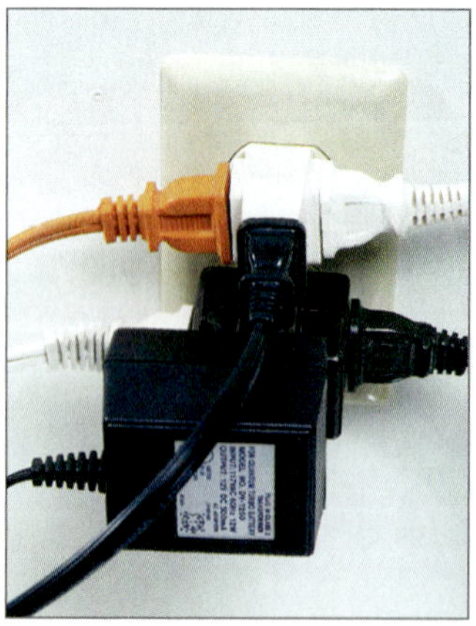

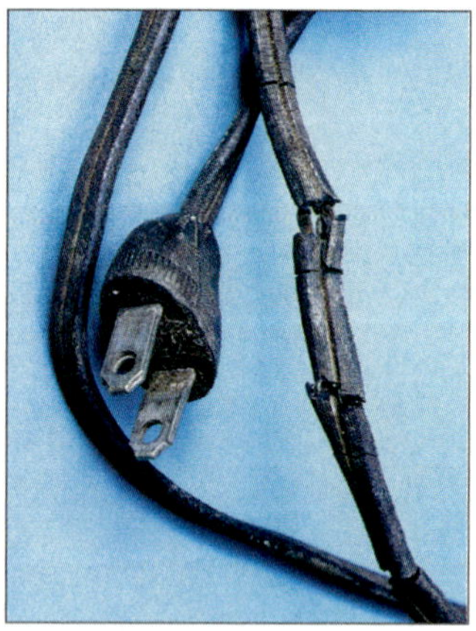

▲ **Figure 5-30** An overloaded circuit or a frayed wire can cause a fire.

1 IDENTIFY: What happens in an overloaded circuit?

Fuses The electricity that comes into your home travels through power lines. The power lines lead to a distribution center. From this center, separate wires go to different circuits in the house or building. These wires are connected to the wall outlets that provide the local power.

Some distribution centers, especially in older homes, have fuses. A **fuse** is a thin piece of metal that melts or "blows" when too much current flows through it. By melting, the fuse stops current from flowing and prevents an overloaded circuit. When you replace the fuse, the current begins moving again. The electrical system in cars and many appliances have fuses to prevent overloaded circuits.

2 DEFINE: What is a fuse?

Circuit Breakers Once a fuse blows, it has to be replaced in order for the current to flow again. In order to avoid having to replace fuses, most newer homes use circuit breakers. A **circuit breaker** is a switch that turns off when there is too much current flowing. A circuit breaker is another way of preventing an overloaded circuit. To start the current flowing again, the breaker must be turned back on.

▲ **Figure 5-31** Circuit breakers prevent overloaded circuits.

3 DESCRIBE: How does a circuit breaker prevent an overloaded circuit?

Electricity in Humans Cells in your body use electricity to carry out very important life processes. For example, the rate at which your heart beats is controlled by electrical signals sent out by certain cells in your body. The muscles that control the movement of your body and your nerves work in a similar manner. This small amount of electricity is produced within your body. It is very dangerous if your body receives an electric current from an outside source. Normal processes within your body can be disrupted by such an electric shock.

4 INFER: List ways in which your body uses electricity.

Electrical Safety Rules Electricity can be dangerous if it is not used carefully. Here are some safety rules to follow when using electricity.

- Do not plug too many appliances into one outlet.
- Do not allow electric cords to become worn.
- Do not run electric cords under carpets.

5 **STATE:** State an electrical safety rule.

CHECKING CONCEPTS

1. Too many appliances connected to one outlet may cause an __________ circuit.
2. Heat and sparks from a frayed wire may cause a __________.
3. Power lines that come into your home lead to a __________.
4. A __________ is a thin piece of metal that melts when too much current moves through it.

THINKING CRITICALLY

5. **HYPOTHESIZE:** Resistance in a wire increases with the length of the wire. Do you think it is a good idea to connect several extension cords to a single outlet? Explain.
6. **ANALYZE:** What can happen if too many appliances are connected to one wall outlet?

HEALTH AND SAFETY TIP

A bolt of lightning can heat the air around it to a temperature hotter than the surface of the Sun! During a lightning storm, if you cannot go indoors, make sure that you stay away from tall buildings and trees. Avoid metal objects, such as baseball bats and even metal umbrellas. Do not go swimming or play in an open field. Why should you not stand under a tree during a lightning storm?

Science and Technology

PACEMAKERS

Heart muscle responds to electrical signals. These signals are used to control the rate of the heart's rhythm. The natural pacemaker of the heart is a small amount of tissue that generates electric charge. This charge spreads through the chambers of the heart to keep them pumping or beating in a repeated rhythm. If this natural pacemaker stops working properly, the normal functions of the body are interrupted. A surgeon may have to insert an artificial pacemaker to keep the heart pumping normally.

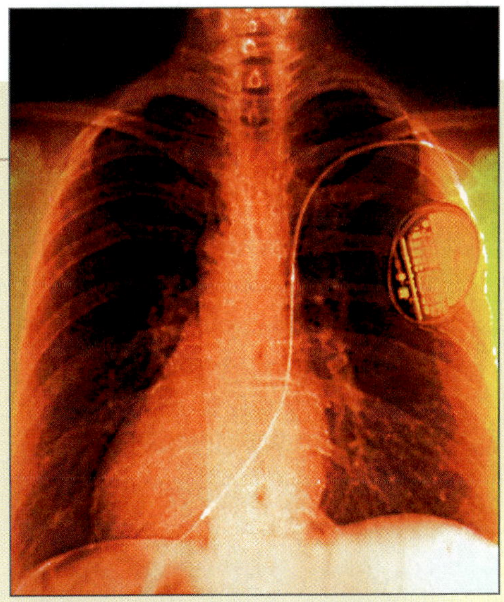

▲ **Figure 5-32** An artificial pacemaker is inserted in the heart.

Artificial pacemakers contain a battery. The battery sends electric charges through a wire inserted into a vein that leads to the heart. The heart muscle responds to these weak electrical charges in the same way that it would to a natural pacemaker. Without electric charges to regulate its pumping rhythm, the heart may beat irregularly. An irregular heartbeat can lead to a heart attack.

Thinking Critically How does an artificial pacemaker work?

Integrating History

THE Big IDEA

How have electrical inventions changed history?

If you have ever switched on a light bulb, listened to recorded music, or watched a movie, you have used a device that runs on electricity. Benjamin Franklin first investigated electricity in the mid 1700s. Now, electricity is a very important source of energy.

The use of electricity has changed history throughout the world. Homes, businesses, and schools all depend on appliances that are powered by electricity. Many of the inventions that use electricity are the products of a famous inventor, Thomas Alva Edison.

Thomas Edison received over 1,000 patents. Some of these patents were for his own inventions, such as the phonograph, the first practical electric light bulb, motion picture technology, and the alkaline storage battery. The other patents were for devices that he improved upon, such as the telephone, the typewriter, the stock ticker, and the printing telegraph.

These inventions, and many others, have had a lasting influence on the world. Because of these many devices that run on electricity, our lives are made easier and more enjoyable. We are able to communicate faster and more easily. We can do more work and perform it faster. We can enjoy music and movies. We can travel safer and faster, and our homes are also filled with electrical devices that make our lives more comfortable.

Learn about some of Thomas Edison's inventions that appear on these two pages. Then follow the directions in the Science Log to come up with your own invention. ✦

Figure 5-33 ▶ Thomas Alva Edison (1847–1931) This American inventor applied for a patent every two weeks during his working career.

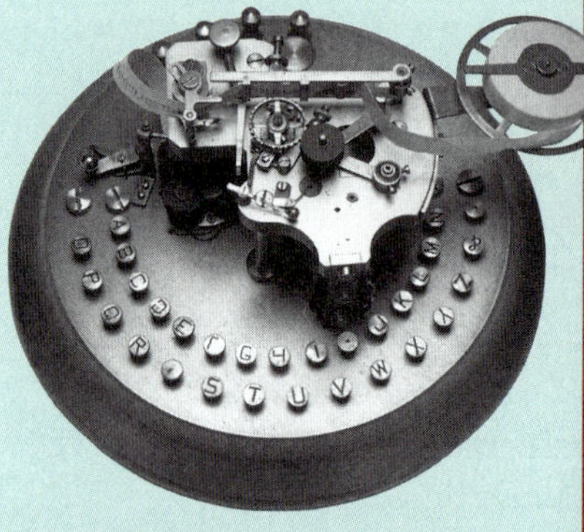

The Printing Telegraph

In 1869, Thomas Edison received a patent for a device that he improved upon, the printing telegraph. This device received messages from great distances and printed them out.

The Phonograph

In 1877, Thomas Edison was the first person to make a machine that recorded sound. This invention led to the development of the phonograph record, and gave birth to the recording industry.

WRITING ACTIVITY

Science Log

Thomas Edison invented devices that he thought were needed by people. Brainstorm about devices that you think people have a need for. In your science log, design one of these devices. Describe it and make labeled drawings. Research how you would have this device patented. Start your search at www.conceptsandchallenges.com.

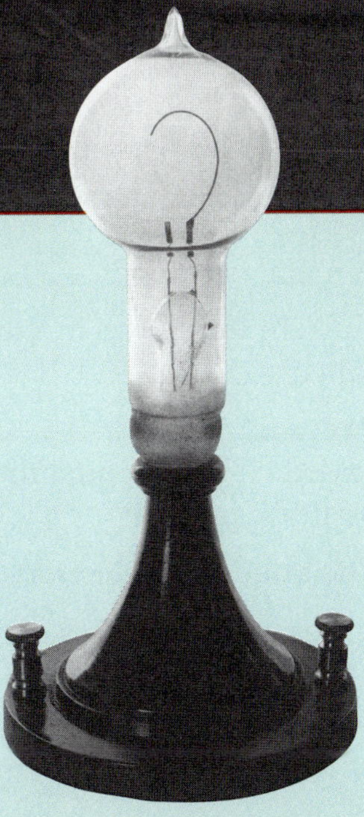

The First Practical Light Bulb

In 1879, with the help of skilled assistants, he tested hundreds of materials to create the first practical light bulb. This light bulb used a filament made of carbon.

The Kinetoscope

In 1892, Edison and his staff completed work on a device that showed moving pictures. The moving pictures were seen through a wooden box called a kinetoscope.

Chapter 5 Challenges

Chapter Summary

Lesson 5-1

- Atoms have three basic parts: **protons, electrons,** and **neutrons.**

Lesson 5-2

- The two types of **electrochemical cells** are a wet cell and a dry cell.

Lesson 5-3

- A **conductor** allows electric charges to flow easily.
- An **insulator** prevents electric charges from flowing through it easily.

Lesson 5-4

- The flow of charges through a conductor is called **electric current.**
- The two types of electric current are **direct current** (DC) and **alternating current** (AC).

Lesson 5-5

- In a **series circuit,** electric current follows only one path.

Lesson 5-6

- In a **parallel circuit,** the electric current can follow more than one path.

Lesson 5-7

- **Volts, amperes,** and **ohms** are the units used to measure **voltage,** current, and **resistance.**

Lesson 5-8

- **Ohm's law** states that current in a wire is equal to the voltage divided by the resistance.

Lesson 5-9

- Electricity must be used safely.

Key Term Challenges

alternating current (p. 120)
ampere (p. 126)
battery (p. 116)
circuit breaker (p. 130)
conductor (p. 118)
direct current (p. 120)
electric circuit (p. 122)
electric current (p. 120)
electrochemical cell (p. 116)
electrode (p. 116)
electrolyte (p. 116)
electron (p. 112)
fuse (p. 130)
insulator (p. 118)
neutron (p. 112)
ohm (p. 126)
Ohm's law (p. 128)
parallel circuit (p. 124)
proton (p. 112)
resistance (p. 126)
series circuit (p. 122)
static electricity (p. 112)
volt (p. 126)

MATCHING Write the Key Term from above that best matches each description.

1. unit for measuring resistance
2. series of electrochemical cells
3. circuit in which the current follows one path
4. electric current that changes direction at a regular rate
5. unit for measuring electric current
6. emergency switch that opens a circuit if too much current is flowing
7. circuit in which electric current can follow more than one path

FILL IN Write the Key Term from above that best completes each statement.

8. A metal is a good __________ of electricity.
9. Current that flows in the same direction is called __________.
10. According to __________, the current in a closed circuit is equal to the voltage divided by the resistance.
11. The buildup of electrons on an object is called __________.
12. The poles of a wet cell are called __________.

Content Challenges TEST PREP

MULTIPLE CHOICE **Write the letter of the term or phrase that best completes each statement.**

1. A battery in a flashlight is an example of
   a. a wet cell.
   b. a dry cell.
   c. alternating current.
   d. a parallel circuit.

2. According to Ohm's law, current is equal to
   a. resistance.
   b. voltage divided by resistance.
   c. resistance multiplied by voltage.
   d. voltage plus resistance.

3. One of the best conductors of electric current is a wire made of
   a. brass.
   b. magnesium.
   c. copper.
   d. rubber.

4. Electric current can follow more than one path in
   a. a series circuit.
   b. an alternating circuit.
   c. a direct circuit.
   d. a parallel circuit.

5. The most common type of current used in homes, offices, and schools is
   a. direct current.
   b. alternating current.
   c. both alternating and direct current.
   d. neither alternating nor direct current.

6. The resistance in a wire will increase if
   a. the thickness of the wire is decreased.
   b. the temperature of the wire is increased.
   c. the length of the wire is increased.
   d. all of the above.

7. If several appliances are connected in series circuit and one of the appliances stops working
   a. none of the appliances will work.
   b. the rest of the appliances will continue to work.
   c. there may be a short circuit.
   d. the current needs to be increased.

8. A circuit breaker is useful in case there is
   a. an overloaded circuit.
   b. not enough current.
   c. not enough voltage.
   d. too much static electricity.

9. Electric current is measured in
   a. volts.
   b. ohms.
   c. amps.
   d. electrons.

10. Most of the circuits in the home are
    a. series circuits.
    b. parallel circuits.
    c. electronic circuits.
    d. short circuits.

TRUE/FALSE **Write *true* if the statement is true. If the statement is false, change the underlined term to make the statement true.**

11. In a dry cell battery, the electrolyte is a dry paste.
12. Current can flow only in a closed circuit.
13. A circuit breaker is used to turn a circuit on or off.
14. In a parallel circuit, the current follows only one path.

Concept Challenges TEST PREP

WRITTEN RESPONSE **Answer each of the following questions in complete sentences.**

1. **HYPOTHESIZE:** During a storm, a power line falls from its pole. Is it safe to touch the wires? Explain.
2. **ANALYZE:** A string of decorative lights goes out when one bulb stops working. Are the lights connected in a series circuit or a parallel circuit? How do you know?
3. **CALCULATE:** Which circuit has more current? **a.** voltage = 12 volts, resistance = 2 ohms **b.** voltage = 32 volts, resistance = 6 ohms **c.** voltage = 24 volts, resistance = 8 ohms
4. **ANALYZE:** How is lightning an example of static discharge?

INTERPRETING VISUALS **Use Figure 5-34 to answer the following questions.**

5. What kind of circuit is shown in the diagram?
6. What are the four parts of this circuit?
7. If you were to connect a light bulb in the circuit, what would happen to the resistance of the circuit?
8. If the voltage was increased to 18 volts, what would be the new value for I?

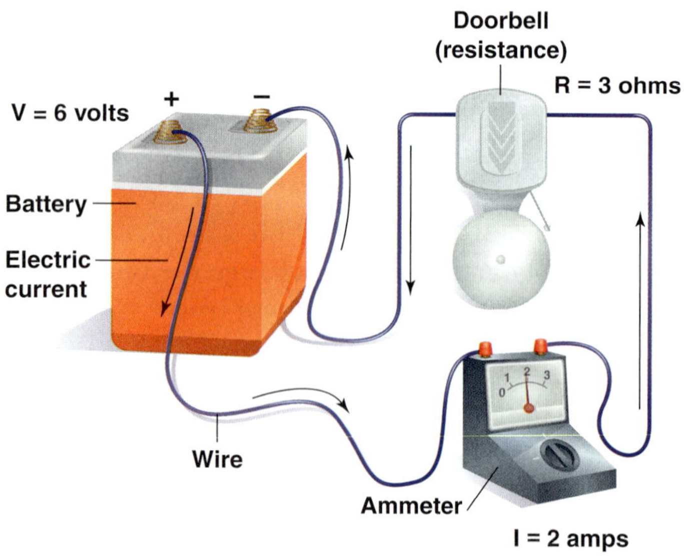

▲ **Figure 5-34** A circuit diagram

Chapter 6 Magnetism

▲ **Figure 6-1** The aurora over the North Pole

The beautiful glow in the sky in Figure 6-1 is not an unusual sight near the North and South poles. The glowing lights are called an aurora. They are caused by particles from the Sun colliding with atoms in Earth's upper atmosphere. Like all objects that have the property of magnetism, Earth has magnetic poles. These magnetic poles produce a magnetic field around Earth. Earth's magnetic field prevents many of the particles from the Sun from entering Earth's atmosphere.

► How does the magnetic field around Earth protect it?

Contents

6-1 What is a magnet?

Objective

Describe the properties of a magnet.

Key Terms

magnetism: force of attraction or repulsion by magnetic materials

pole: end of a magnet

Magnetism When you hold a magnet close to certain types of metals, the metals move toward the magnet. The metals are pulled toward the magnet by a force called magnetism. **Magnetism** is a force of attraction or repulsion by magnetic materials.

Not all matter is attracted to a magnet. For example, if you hold a magnet near a piece of wood, the wood will not be attracted to the magnet. Copper, aluminum, plastics, and glass are other materials that are not attracted to a magnet. Elements that are attracted to a magnet include iron, nickel, and cobalt. Objects made of steel, which is mostly iron, are also attracted to magnets.

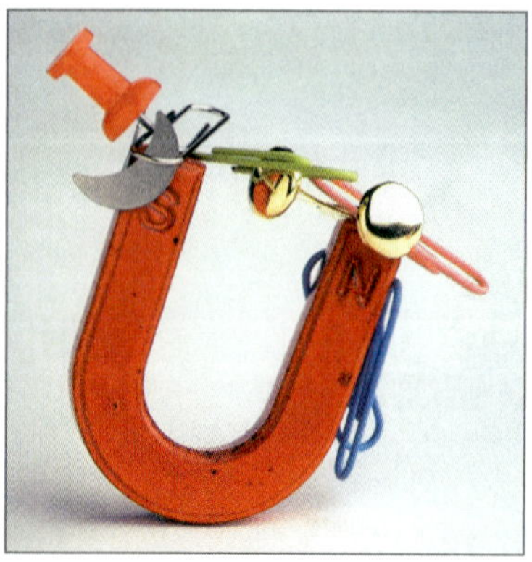

◄ **Figure 6-2** Materials such as iron and steel are attracted to magnets.

1 DEFINE: What is magnetism?

Magnetic Poles Each end of a magnet is called a magnetic **pole.** The magnetic forces of a magnet are strongest at its poles. Every magnet has two poles—a north pole (N) and a south pole (S). If you hang a magnet by a string, it will turn until its north pole points north and its south pole points south. Thus, the north pole of a magnet is sometimes called the north-seeking pole and the south pole of a magnet is sometimes called the south-seeking pole. Magnets always have two poles. Even if a magnet were broken up into tiny pieces, each tiny piece would still have a north pole and a south pole.

2 DEFINE: What is a pole?

Properties of a Magnet Magnetic poles behave in the same way that electric charges do when like and unlike charges are brought together. If a north pole of one magnet and a south pole of another magnet are brought together, they will attract each other. The unlike poles of any two magnets always attract each other in the same way that unlike electric charges attract each other.

If the north pole of one magnet is brought near the north pole of another magnet, they will repel, or push away from, each other. Two magnetic south poles will also repel each other. The like poles of any two magnets always repel each other in the same way that like electric charges repel each other.

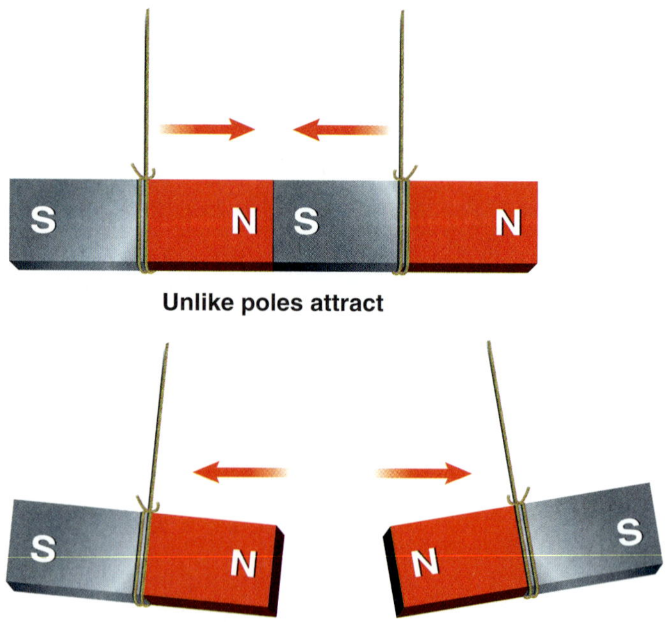

▲ **Figure 6-3** How magnets behave

3 DESCRIBE: How do the poles of a magnet affect each other?

Uses of Magnets You may not realize it, but magnets play a very important role in your life. They are used in electric can openers, telephones, doorbells, stereo speakers, electric motors, and alarm systems. Magnets are also used in high-energy particle research and to run special high-speed trains.

4 LIST: Name some ways in which magnets are used.

CHECKING CONCEPTS

1. Magnetism is a force of ___________ or repulsion by magnetic materials.
2. When hung from a string, the north pole of a magnet points ___________.
3. The south pole of a magnet is sometimes called the ___________.
4. The north pole of one magnet will ___________ the south pole of another magnet.
5. Elements attracted to a magnet are iron, ___________, and cobalt.
6. If you cut a magnet in half, each half will have a north pole and a ___________.

THINKING CRITICALLY

7. INFER: Why could the north pole of a magnet be called the north-seeking pole?
8. DESCRIBE: Describe what will happen if you bring the north pole of one magnet near the north pole of another magnet.
9. ANALYZE: How are magnets similar to electric charges?
10. INFER: How do you think a magnet is used in an electric can opener?

DESIGNING AN EXPERIMENT

Design an experiment to solve the following problem. Include a hypothesis, variables, a procedure, and a type of data to study.

PROBLEM: A homeowner has a door that she wants to keep slightly open. How can magnets be used to keep the door from shutting?

How Do They Know That?

MAGNETIC COMPASS

Magnets can be found in nature. Naturally occurring substances with magnetic properties are called natural magnets. The ancient Greeks found natural deposits of magnetic iron ore. They discovered this ore in a region of Turkey known as Magnesia. The ore is called magnetite (MAG-nuh-tyt). Magnetite can also be found in many other parts of the world.

▲ **Figure 6-4** Sailors once used a magnetic compass like this one.

Magnetite was used by sailors and navigators to find directions at sea. Pieces of magnetite were called lodestone because sailors found that a piece of magnetite always pointed toward the North Star. The North Star was called the lodestar, or leading star. Sailors made magnetic compasses with a pointer made of lodestone.

Today, magnetic compasses are often used by hikers and backpackers. The pointer of a magnetic compass always points north. Once you know where north is, you can easily locate the other three directions, or compass points.

Thinking Critically If a hiker is trying to hike in an easterly direction, would a compass be of any help to the hiker? Explain your answer.

6-2 What causes magnetism?

INVESTIGATE

Finding the Pole of a Magnet
HANDS-ON ACTIVITY

1. Cover the pole markings of a bar magnet with masking tape. Tie one end of a piece of string around the center of the magnet. Tape the other end of the string to the edge of your desk or table. Allow the magnet to hang from your desk or table.
2. Use another bar magnet to determine the poles of the hanging magnet.

THINK ABOUT IT: How did you determine the poles of the hanging magnet?

Objective
Demonstrate the shape of a magnetic field.

Key Terms
magnetic field: describes the region of magnetic force around a magnet

magnetic lines of force: lines that show the shape of a magnetic field

magnetic domain: groups of atoms whose magnetic poles are all lined up in the same direction

Magnetic Field If you bring a magnet close to an iron object, the magnetic force will attract the object. The magnet does not need to touch the object in order for the magnetic force to attract it. This attraction happens because every magnet has a magnetic field around it. The **magnetic field** describes the region of magnetic force around a magnet.

The magnetic fields around different magnets are not the same. Some magnets may have a stronger magnetic field than others do. Figure 6-5 shows the magnetic field around a bar magnet. Iron filings have been sprinkled over the bar magnet. The iron filings line up around the magnet showing the shape of the magnet's magnetic field.

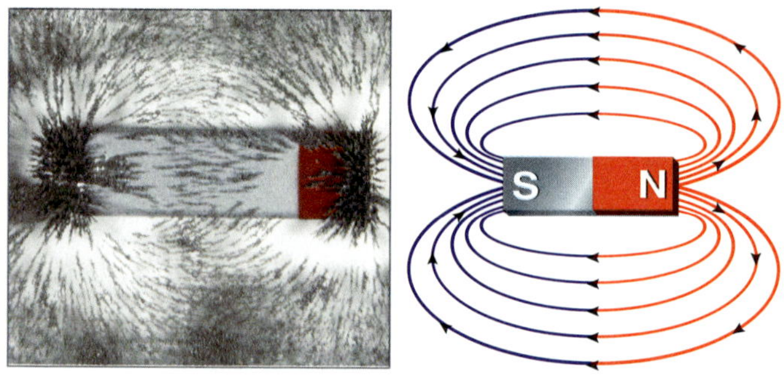

▲ **Figure 6-5** Iron filings sprinkled on plastic covering a bar magnet line up along the magnetic lines of force around the magnet.

1 DEFINE: What is a magnetic field?

Magnetic Lines of Force A magnetic field is made up of **magnetic lines of force.** When iron filings are sprinkled around a magnet, they will line up along the magnetic lines of force. Magnetic lines of force always begin at one pole and end at the opposite pole. They are also closest together at the poles because the magnetic field is strongest at the poles.

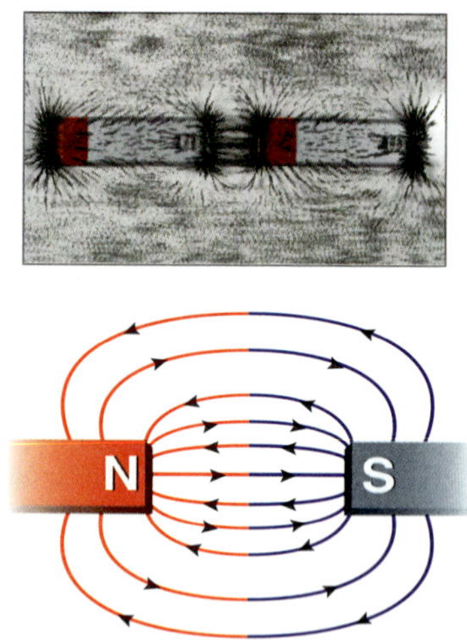

▲ **Figure 6-6** Magnetic lines of force show that unlike poles attract.

The magnetic lines of force between two like poles can be seen in Figure 6-7. The lines of force show that two like poles repel each other. The lines of force from the north pole of each magnet are bent away from the north pole of the other magnet.

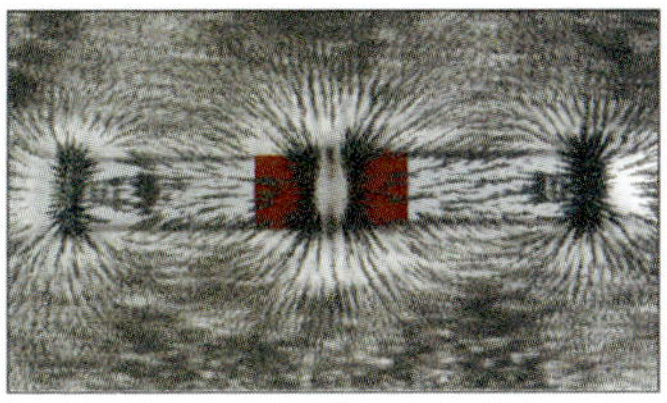

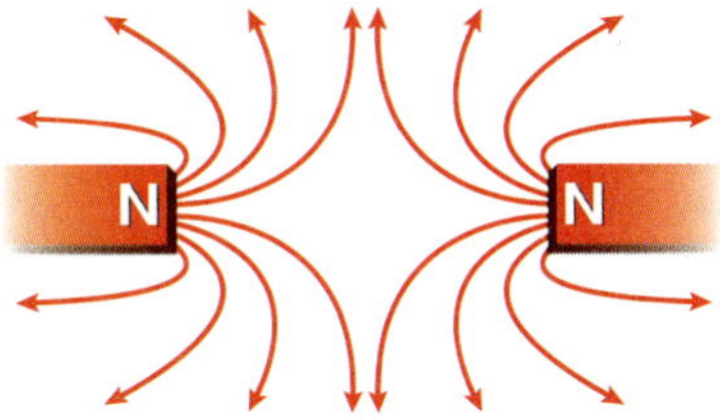

▲ **Figure 6-7** Magnetic lines of force show that like poles repel.

Two unlike poles attract each other. The lines of force from the north pole of one magnet bend toward the south pole of the other magnet.

2 **DESCRIBE:** What do the magnetic lines of force show?

Magnetic Domains All matter is made up of atoms. The motion of electrons causes all atoms to have magnetic properties. In magnetic materials such as iron, groups of atoms can join together so that the magnetic fields of each atom point in the same direction. The north pole of each atom all point in the same direction. The south pole of each atom all point in the opposite direction. Groups of atoms whose magnetic poles are all lined up in the same direction are called **magnetic domains.**

Most of the magnetic domains of magnetic materials point in the same direction. However, the domains of unmagnetized materials point in different directions. Because the domains point in different directions, their magnetic fields cancel each other.

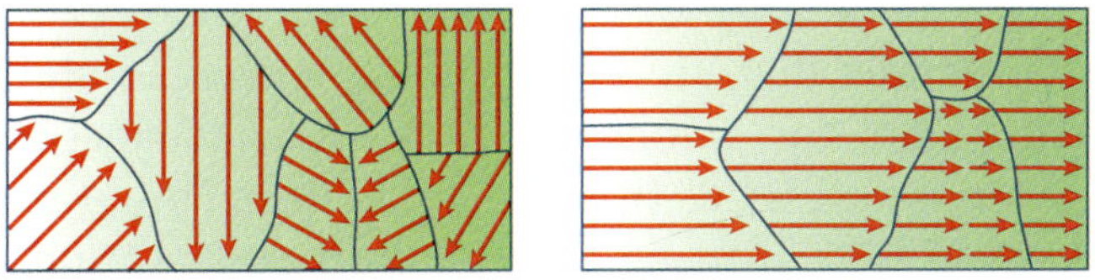

Unmagnetized material **Magnetized material**

▲ **Figure 6-8** The domains of unmagnetized materials point in different directions. The domains of magnetized materials point in the same direction.

3 **DEFINE:** What is a magnetic domain?

CHECKING CONCEPTS

1. What is a magnetic field?
2. What happens when you bring a magnet near a piece of iron?
3. What are magnetic lines of force?
4. How can iron filings be used to show magnetic lines of force?
5. Why are magnetic lines of force closest together at the poles of a magnet?
6. What causes all atoms to have magnetic properties?

THINKING CRITICALLY

7. **PREDICT:** What will happen to the magnetic lines of force when the north poles of two magnets are placed next to each other?
8. **EXPLAIN:** How can magnetic lines of force show that unlike poles attract each other?
9. **COMPARE:** How are the domains of magnetized and unmagnetized materials different?
10. **INFER:** If an object made of iron is brought near a magnet without actually touching it, how can the magnetic force of the magnet attract the object?

INTERPRETING VISUALS

Use Figure 6-9 to answer the question.

11. **ANALYZE:** Is pole *A* a north pole or a south pole? How do you know?

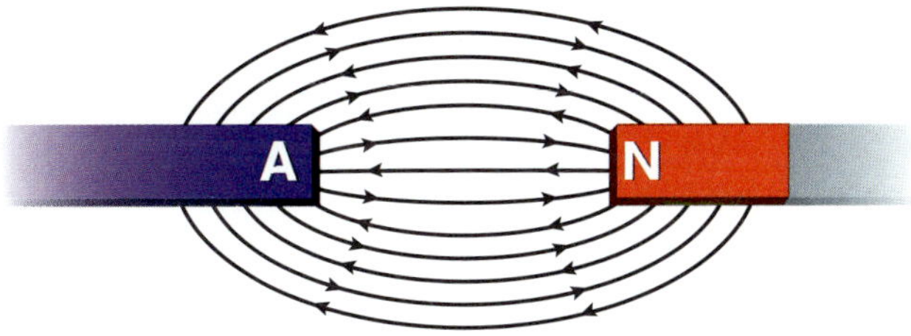

▲ **Figure 6-9** Magnetic lines of force around two magnets

6-3 How can you make a magnet?

Objectives

Explain how materials can be magnetized.
Compare permanent magnets and temporary magnets.

Key Term

magnetic induction: process by which a material can be made into a magnet

Natural Magnets Some magnets occur in nature. These magnets are called natural magnets. Magnetite, or lodestone, is an example of a natural magnet. Certain materials that are not natural magnets can be made into magnets, or magnetized (MAG-nuh-tyzd). For example, rubbing an iron nail with a magnet in one direction will magnetize the nail. The magnetized nail will have the same properties as a natural magnet.

▲ **Figure 6-10** Magnetite, or lodestone, is a natural magnet.

1 NAME: What is an example of a natural magnet?

Magnetic Induction The process by which a material is magnetized, or made into a magnet, is called **magnetic induction.** In magnetic induction, the domains of certain materials are made to point in the same direction. They become magnetized. Some materials are easier to magnetize than others. Alloys of iron, nickel, and cobalt are easy to magnetize. Rubbing objects made of these materials with a magnet can cause the domains within the objects to point in the same direction. When this happens, these materials become magnetized. Other materials show weak magnetic properties. For example, if an aluminum object is placed in the magnetic field of a strong magnet, the atoms of the metal may be weakly affected by the magnetism. However, when the object is removed from the magnetic field, the domains will go back to pointing in different directions.

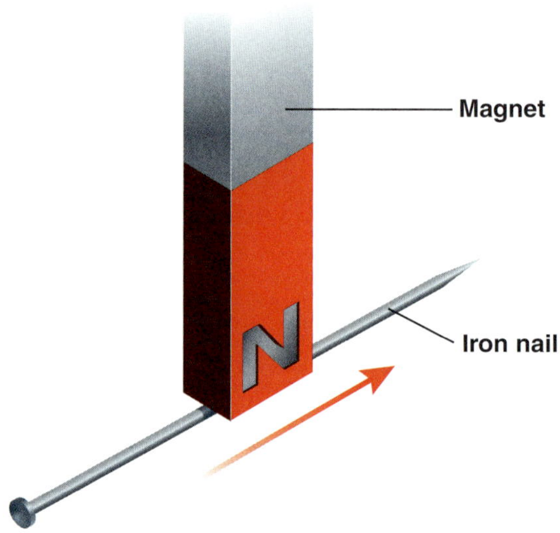

▲ **Figure 6-11** Magnetic induction causes the iron nail to become magnetized.

2 DEFINE: What is magnetic induction?

Temporary and Permanent Magnets Once a material is magnetized, it may or may not remain magnetized. Iron is easily magnetized, but the iron loses its magnetism quickly. A material that is easily magnetized tends to lose its magnetism quickly. A magnet made of this kind of material is called a temporary magnet.

Materials that are hard to magnetize can stay magnetized for a long time. A piece of steel is harder to magnetize than a piece of iron. However, the steel will keep its magnetic properties much longer than will a piece of iron. A magnet that is hard to magnetize but tends to keep its magnetism is called a permanent magnet.

3 CLASSIFY: Is an iron magnet a temporary magnet or a permanent magnet?

Losing Magnetism All magnets, even permanent magnets, can lose their magnetism. If you drop a magnet, you can cause the magnetic domains inside of it to stop pointing in the same direction. When the magnetic domains point in different directions, the magnet loses some or all of its magnetic properties. Heating a magnet can also cause it to become demagnetized. The energy from the heat can cause the atoms to move faster and farther apart. When this happens, the magnetic domains are no longer aligned.

4 **EXPLAIN:** What are two ways in which a magnet can become demagnetized?

CHECKING CONCEPTS

1. An example of a natural magnet is ___________.
2. Rubbing an iron nail in the same direction with a magnet will ___________ the nail.
3. When a piece of iron becomes magnetized, the magnetic ___________ point in the same direction.
4. A ___________ magnet is hard to magnetize.
5. A ___________ magnet will not keep its magnetic properties for a long time.
6. All magnets can lose their ___________.
7. Dropping a magnet can cause its ___________ to point in different directions.

THINKING CRITICALLY

8. **INFER:** An iron nail has been magnetized. Does the magnetized nail have a north pole and a south pole? Explain.
9. **CONTRAST:** What is the difference between a temporary magnet and a permanent magnet?
10. **PREDICT:** Soft iron is very easy to magnetize. Will soft iron make a good permanent magnet? Why or why not?
11. **EXPLAIN:** How does dropping a magnet affect the magnetic domains within the magnet?

BUILDING SOCIAL STUDIES SKILLS

Researching The properties of the ore magnetite have been known for more than 2,000 years. The ancient Chinese may have discovered the properties of magnetite as early as 2600 B.C. Use library references, encyclopedias, and the Internet to find out how the ancient Chinese used magnetite and how they explained its properties in stories and myths.

Hands-On Activity

MAKING A MAGNETIC COMPASS

▲ **STEP 4** Observe in which direction the needle points.

You will need a sewing needle, a bar magnet, a pushpin, a cork, a small plastic container of water, and a magnetic compass.

1. Insert the pushpin into the bottom of the cork.
2. Float the cork, with the pushpin pointing down, in a container of water.
3. Rub the magnet against the needle about 20 times. Rub in one direction only.
4. Lay the needle on top of the cork. Observe in which directions the needle points when it stops turning. Using a magnetic compass away from your setup, compare the direction of the needle with the direction of the compass pointer.

Practicing Your Skills

5. **DESCRIBE:** What effect does rubbing with a magnet have on the needle?
6. **OBSERVE:** In which directions does the floating needle point?

6-4 How is Earth like a magnet?

Objective

Explain how Earth acts like a magnet.

Key Term

magnetosphere (mag-NEET-oh-sfeer)**:** region of Earth's magnetic field

William Gilbert In the early 1600s, British scientist William Gilbert made several observations about magnetism. Gilbert observed that if you hang a bar magnet from a string, one pole of the magnet always points north and the other pole always points south. William Gilbert was the first person to suggest an explanation for this observation. He proposed that Earth itself is a magnet, and just like any magnet, Earth has a north pole and a south pole.

1 STATE: How did Gilbert's proposal compare Earth to a magnet?

Earth's Magnetic Poles Scientists have discovered that William Gilbert's proposal was correct. Earth does act as if a huge bar magnet is buried deep inside of it. This causes Earth to have two magnetic poles. One of Earth's magnetic poles is near the geographic North Pole in Canada. The other magnetic pole is near the geographic South Pole in Antarctica. The magnetic poles of Earth are not exactly at Earth's geographic poles. In fact, the magnetic pole near Earth's geographic North Pole is about 1,300 kilometers away from it.

2 EXPLAIN: Where are Earth's magnetic poles in relation to Earth's geographic poles?

Earth's Magnetic Poles Are Not Earth's Geographic Poles You have learned that the north pole of a magnet always points north. However, like magnetic poles repel each other. This means that when a magnet points to the geographic North Pole, the North Pole is really Earth's magnetic south pole. It also means that when the south pole of a magnet faces Earth's geographic South Pole, the South Pole must be Earth's magnetic north pole. Figure 6-12 shows Earth's magnetic poles and geographic poles.

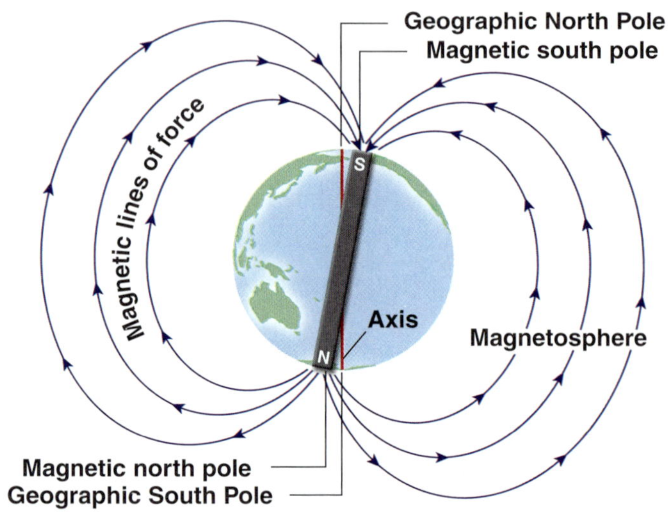

▲ **Figure 6-12** Earth's magnetic poles are not the same as Earth's geographic poles.

3 INFER: Why is Earth's geographic North Pole called the magnetic south pole?

The Magnetosphere Like any magnet, Earth is surrounded by a magnetic field. The magnetic field is strongest near the north and south magnetic poles. Earth's magnetic field extends far into space. The magnetic field around Earth is called the **magnetosphere.**

The magnetosphere traps charged particles released by the Sun. Most of these particles are deflected by the magnetosphere. However, some particles do get through it. When these charged particles hit Earth's atmosphere, they interact with other particles, causing the air above Earth's surface to glow. This glowing area is called an aurora (aw-RAWR-uh).

Auroras seen in the Northern Hemisphere near the North Pole are also called the northern lights or aurora borealis. Auroras seen in the Southern Hemisphere near the South Pole are called the southern lights, or aurora australis.

▲ **Figure 6-13** The southern lights are caused by particles from the Sun interacting with particles in Earth's atmosphere.

4 **DEFINE:** What is the magnetosphere?

CHECKING CONCEPTS

1. When a compass points north, it is pointing to Earth's magnetic __________ pole.
2. Earth's magnetic south pole attracts the __________ pole of a magnet.
3. Earth's geographic North Pole is also Earth's magnetic __________ pole.
4. Earth is surrounded by a magnetic field called the __________.

THINKING CRITICALLY

5. **CLASSIFY:** Is Earth a permanent magnet or a temporary magnet? How do you know?
6. **INFER:** Explain why auroras occur in both the Northern Hemisphere and the Southern Hemisphere.

BUILDING SCIENCE SKILLS

Making a Model The Sun produces a stream of high-energy particles called solar wind. Solar wind causes the magnetosphere to form into a special shape. Find out the shape of the magnetosphere. Then, use markers and poster board to make a model of it. Share your illustration with your class.

How Do They Know That?

THE HISTORY OF EARTH'S MAGNETISM IN ROCKS

Scientists have found that Earth acts like a giant magnet. The magnetism comes from deep within the planet. Earth's center, or core, is surrounded by hot liquid metal. As the liquid metal flows, it produces a magnetic field. The flowing liquid metal around the core changes over time. These changes cause Earth's magnetic field to change with it. In fact, every few thousand years, the magnetic field of the Earth reverses itself. Geologists discovered this fact by studying the magnetic fields in certain rocks.

▲ **Figure 6-14** Volcanic rock contains magnetic stripes that can be detected by special instruments.

Rocks containing the metal iron can become permanent magnets when they form. These magnetic rocks can form inside volcanoes. As lava from the volcano cools, Earth's magnetic field pulls on the magnetic materials in the rock. The magnetic materials in the rock line up to match the magnetic field of Earth at the time that the rock formed.

Geologists study the magnetic patterns inside these rocks. Special instruments are used to study these rocks because the magnetic patterns cannot be seen with the naked eye. From the magnetic patterns in many of the rocks, scientists have learned about the reversals of Earth's magnetic field. A reversal of Earth's magnetic field occurs every few thousand years.

Thinking Critically How can magnetic rocks reveal that Earth's magnetic field reverses itself?

6-5 How are electricity and magnetism related?

INVESTIGATE

Observing Magnetism and Electricity
HANDS-ON ACTIVITY

1. Note in what direction the pointer of a magnetic compass points.
2. Connect two insulated wires to a dry cell (size D) battery and a light bulb in a holder.
3. When the light bulb lights up, carefully bring the compass near the wires. Observe the pointer of the compass.

THINK ABOUT IT: What happens to the pointer of the compass?

Objective
Explain the relationship between electricity and magnetism.

Key Terms
electromagnetism: relationship between electricity and magnetism

electromagnetic induction: process by which an electric current is produced by moving a wire in a magnetic field

Oersted's Discovery Have you ever used a magnetic compass around an electrical device? If you have, you may have noticed that the pointer of the compass gets "confused." The pointer may no longer point north. This happens because of the relationship between electricity and magnetism. This relationship is called **electromagnetism.** The Danish scientist Hans Christian Oersted (UR-sted) was the first person to show that this relationship does exist.

More than 150 years ago, Hans Christian Oersted discovered an important property of electric current. Oersted connected a simple series circuit made up of a battery and a wire. As charges flowed through the wire, Oersted noticed that a nearby compass pointer moved. When the circuit was disconnected, the compass pointer returned to its original position. He also noticed that when the direction of the current was reversed, the pointer moved in the opposite direction. Hans Christian Oersted discovered that an electric current produces a magnetic field.

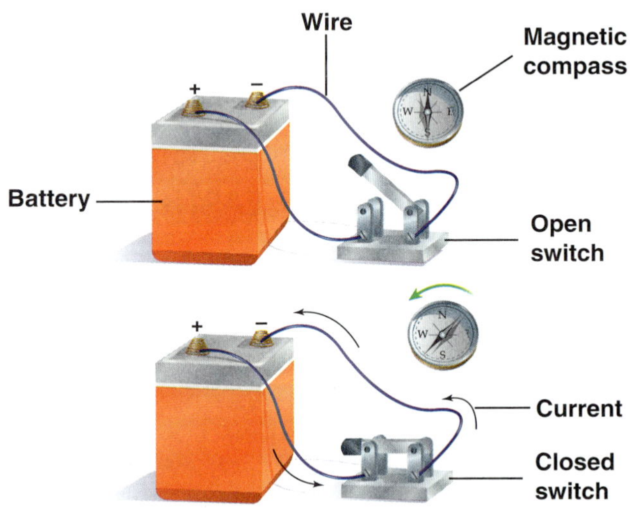

▲ **Figure 6-15** When electric current flows through a circuit, a magnetic field is produced.

1 DESCRIBE: What did Hans Christian Oersted discover?

Electromagnetic Induction An electric current produces a magnetic field. Does a magnetic field produce an electric current? A British scientist named Michael Faraday tried to answer this question. He hung a wire that remained still. Then he moved a strong magnet around it. He found that when the magnet was held still, no electric current moved through the wire. However, when he moved the magnet toward and away from the wire continuously, electric charges flowed through the wire. He also noticed that the direction of the current depended upon the direction of the movement of the magnet.

Faraday also tried moving a wire coil around a stationary magnet and got the same results. Thus, Faraday found that electric current will be produced in a wire that is exposed to a changing magnetic field. The process by which an electric current is produced when a wire is exposed to a changing magnetic field is called **electromagnetic induction.** The current is induced, or caused to flow, in a wire when the wire cuts across magnetic lines of force.

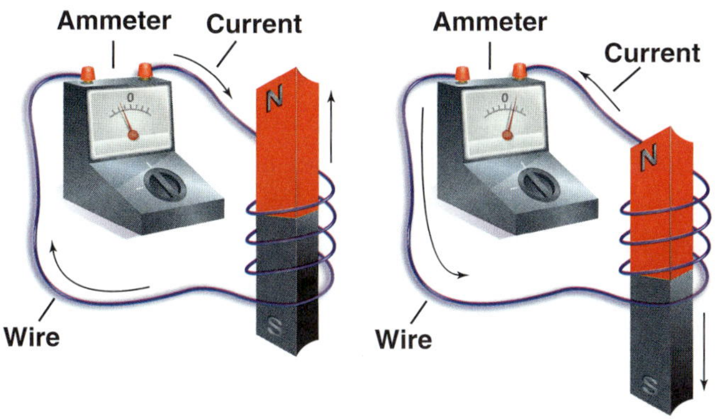

▲ **Figure 6-16** In electromagnetic induction, electric current is produced in a wire by a changing magnetic field.

2 **DEFINE:** What is electromagnetic induction?

CHECKING CONCEPTS

1. Oersted discovered that an __________ caused a magnetic field.
2. An electric current is produced when a wire is exposed to a __________ magnetic field.
3. Michael Faraday discovered electromagnetic __________.
4. Electromagnetism is the relationship between __________ and magnetism.

THINKING CRITICALLY

5. **DESCRIBE:** What determines the direction of electric current in a wire when a magnet is moved around the wire?
6. **EXPLAIN:** What happens to a nearby compass when you disconnect a circuit?
7. **EXPLAIN:** Why will an electric current not be produced in a wire that is exposed to a stationary magnetic field?

People in Science

JOSEPH HENRY (1797–1878)

▲ **Figure 6-17** Joseph Henry

Joseph Henry was an American scientist. He studied electricity and magnetism. He and the British scientist Michael Faraday made similar discoveries at about the same time. Joseph Henry discovered that a changing magnetic field will induce a current in a conductor. However, Faraday is given credit for making the same discovery.

Joseph Henry was born in Albany, New York, on December 17, 1797. He helped Samuel Morse develop the telegraph. In 1831, Henry built a telegraph of his own. He also built one of the most powerful devices known at the time. Joseph Henry's device used electromagnetism to lift more than a ton of weight.

In 1846, Joseph Henry was appointed to head the Smithsonian Institution in Washington, D.C. He was also the main organizer of the National Academy of Science and started the U.S. Weather Service. In honor of Joseph Henry, the unit for electromagnetic induction is called the henry.

Thinking Critically Why is the unit for electromagnetism called the henry?

6-6 What is an electromagnet?

Objective

Describe how to make an electromagnet.

Key Term

electromagnet: temporary magnet made by wrapping a current-carrying wire around a metal core

Magnetic Field Strength A wire carrying an electric current always has a magnetic field around it. The magnetic field around a straight wire is not very strong. If the wire is wound into a coil, the magnetic field becomes much stronger. The magnetic fields around each coil of wire add together. The more turns you add to the coil, the stronger the magnetic field will become. Increasing the amount of current running through the wire will also increase the magnetic field around the wire.

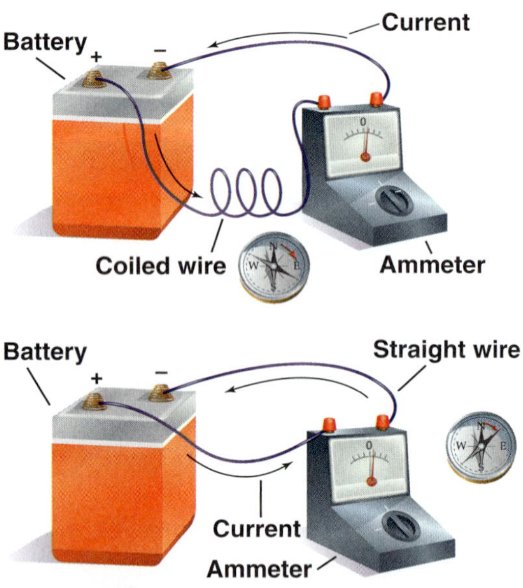

▲ **Figure 6-18** The compasses show that strength of the magnetic field around a coiled wire is stronger than the strength of the magnetic field around a straight wire.

1 **PREDICT:** Which will have a stronger magnetic field, a wire coil with two turns or a wire coil with four turns?

Electromagnets The strength of the magnetic field around a wire can be increased by adding turns to the wire coil and by increasing the amount of current flowing through the wire. However, wrapping coils of wire around a piece of metal, such as iron, will produce the greatest strength in the magnetic field around the wire. The magnetic field around the wire aligns the magnetic domains within the piece of iron. The strength of the magnetic field around the coiled wire and the strength of the magnetic field around the iron add together.

When electric current flows through a coil of wire that is wrapped around an iron core, a strong magnet is produced. This type of magnet is called an **electromagnet.** The magnetic material in the center of an electromagnet is called a core. An electromagnet is a temporary magnet. As long as current is flowing through the wire, the electromagnet has a strong magnetic field. When the current is turned off, there is no longer a magnetic field.

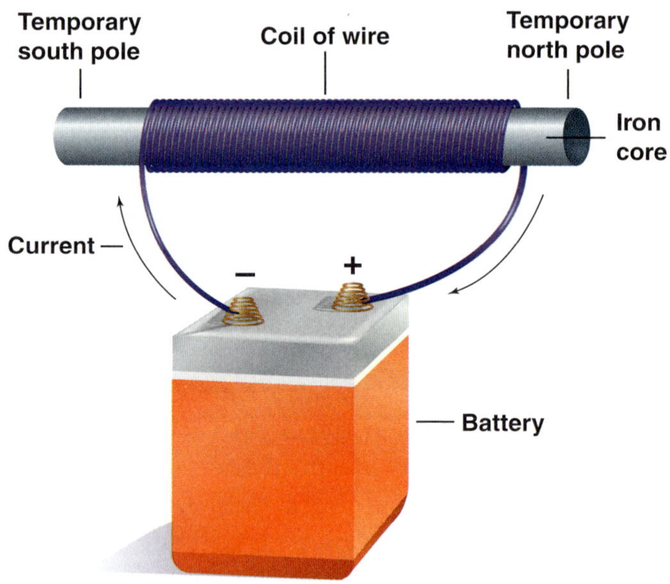

▲ **Figure 6-19** An electromagnet is made up of an energy source, a wire coil, and a metal core.

The strength of an electromagnet can be increased in two ways. Adding more turns to the wire coil around the magnetic core makes an electromagnet stronger. Increasing the current through the wire also makes an electromagnet stronger.

2 **INFER:** What are two ways to make an electromagnet stronger?

Uses of Electromagnets Electromagnets have many important uses. They are used in devices, such as radios, telephones, and computers. Large electromagnets are used to lift heavy pieces of metal. Electromagnets are very useful because they can be turned on and off.

▲ **Figure 6-20** Electromagnets can be used to lift heavy objects.

3 **EXPLAIN:** Why are electromagnets so useful?

CHECKING CONCEPTS

1. What is an electromagnet?
2. Why is an electromagnet considered to be a temporary magnet?
3. What are two ways to increase the strength of the magnetic field around a wire?
4. Name three devices that use electromagnets.

THINKING CRITICALLY

5. **CONTRAST:** How is an electromagnet different from a permanent magnet?
6. **LIST:** What are the materials needed to make an electromagnet?

INTERPRETING VISUALS

Look at Figure 6-19 of an electromagnet to answer the following questions.

7. **PREDICT:** What would happen if you were to disconnect the wires from the dry cell?
8. **ANALYZE:** What would happen if there were fewer coils wrapped around the metal core?

Hands-On Activity

MAKING AN ELECTROMAGNET

You will need paper clips, a D-cell battery, a 6-volt battery, insulated wire, and an ungalvanized nail.

1. Wrap 20 turns of wire around an ungalvanized nail. Connect the ends of the wire to the terminals of the D-cell battery.
2. Try to pick up some paper clips.
3. Now, wrap 40 turns of wire around the nail. Connect the ends of the wire to the same battery. See how many paper clips you can pick up.
4. Connect the electromagnet with 40 turns of wire to the 6-volt battery. See how many paper clips you can pick up.

▲ **STEP 4** Pick up paper clips with your electromagnet.

Practicing Your Skills

5. **OBSERVE:** How many paper clips did the electromagnet with 20 turns of wire connected to the D-cell battery pick up? With 40 turns?
6. **COMPARE:** Were you able to pick up more paper clips when the electromagnet was connected to the D-cell battery or to the 6-volt battery? Explain.

6-7 What is a transformer?

Objectives

Explain how a transformer works. Compare a step-up and a step-down transformer.

Key Term

transformer: device in which alternating current in one coil of wire induces a current in a second coil of wire

Transformers Suppose you wrap two coils of wire around a nail. You attach one of the coils to a dry cell and a switch. This coil is called the primary coil. You connect the other coil to an ammeter. This is the secondary coil. When you close the switch in the primary coil, the ammeter shows that a current flows in the secondary coil. This current quickly dies out. When you open the switch, current again flows in the secondary coil. Again, it quickly dies out. To keep current flowing to the secondary coil, you must keep opening and closing the switch.

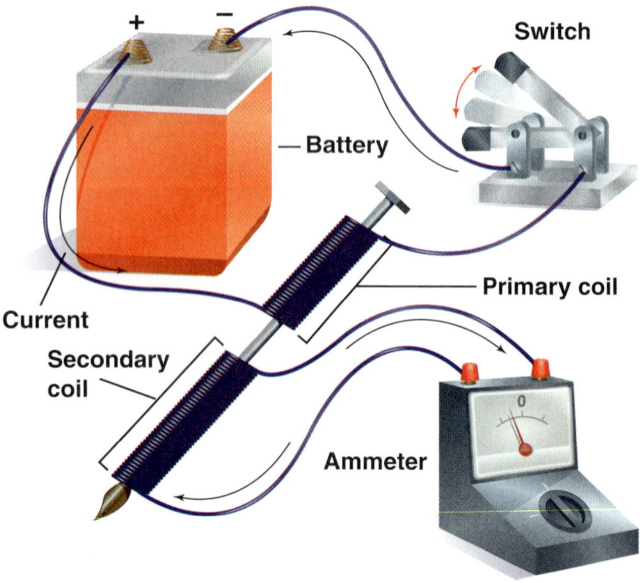

▲ **Figure 6-21** To keep current flowing to the secondary coil, the switch has to be opened and closed over and over again.

A **transformer** is a device that uses alternating current in the primary coil to induce a current in the secondary coil. The primary coil is connected to a source of alternating current.

1 DEFINE: What is a transformer?

Step-Up Transformers A transformer can be used to increase the voltage of alternating current. This kind of transformer is called a step-up transformer. A step-up transformer has more turns of wire in the secondary coil than in the primary coil. The more turns of wire there are in the secondary coil, the higher the voltage will be. Power companies use step-up transformers to produce high-voltage electricity to send over long distances.

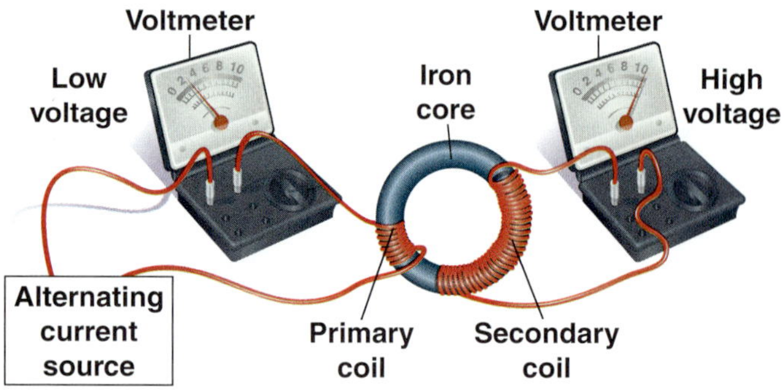

▲ **Figure 6-22** A step-up transformer has fewer coils in the primary coil than in the secondary coil.

2 DESCRIBE: What does a step-up transformer do?

Step-Down Transformers A transformer can also decrease the voltage of alternating current. A transformer that decreases voltage is called a step-down transformer. A step-down transformer has fewer turns of wire in the secondary coil than in the primary coil. The fewer the turns of wire in the secondary coil, the lower the voltage will be. Step-down transformers lower the high voltage carried by power lines.

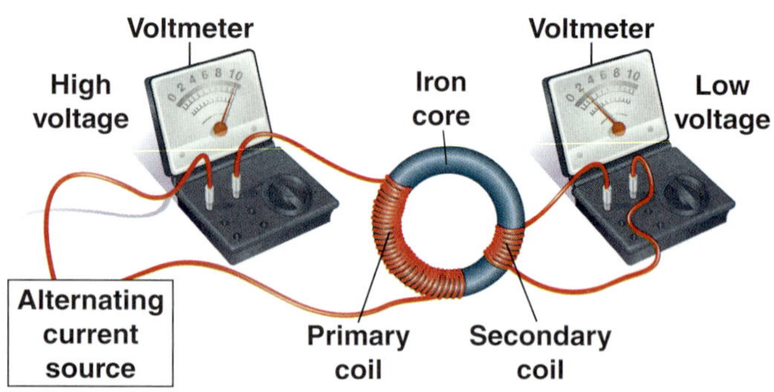

▲ **Figure 6-23** A step-down transformer has more coils in the primary coil than in the secondary coil.

3 EXPLAIN: What does a step-down transformer do?

Transmission of Electricity Power plants use transformers to send electricity to homes, businesses, and schools. Very often, power plants are located in areas far away from where electricity is needed. Electricity has to travel long distances to reach these areas. At power plants, electricity is produced at low voltage. However, when electricity travels long distances, there is a loss of energy. By using a step-up transformer, power companies are able to send high-voltage electricity over long distances without too much loss of energy. Because electricity is usually used at a lower voltage, step-down transformers lower the voltage before the electricity is sent into homes, businesses, and schools.

4 **INFER:** When is a step-down transformer used?

CHECKING CONCEPTS

1. What is a transformer?
2. Which coil of a transformer is attached to a source of current?
3. What kind of transformer can be used to increase voltage?
4. What kind of transformer can be used to decrease voltage?

THINKING CRITICALLY

5. **CONTRAST:** What is the difference between a step-up transformer and a step-down transformer?

BUILDING SCIENCE SKILLS

Investigating The voltage from a wall outlet in your home is 120 volts. Unplug and look at several appliances around your home. How much voltage do they need to operate? The voltage is usually given on the appliance or in the owner's manual. Which appliances need less than 120 volts? Which need more than 120 volts? Do the appliances use step-down or step-up transformers?

Real-Life Science

BIRDS ON A WIRE

▲ **Figure 6-24** Birds do not form a circuit with the electricity in a power line.

High-voltage electricity is transmitted through electric power lines. The voltage can range from about 7,200 to 800,000 volts. Electricity that escapes from these lines can do great damage. Sparks from fallen lines can cause fires.

You should never go near power lines. You can be seriously injured if enough electricity passes through your body. Power workers who repair power lines have to wear protective clothing when they work on the lines. Yet, birds can sit on power lines all day long without getting hurt. Why are birds not in danger when they sit on electric power lines?

The answer is that birds do not form a circuit with the electricity in the power lines. Because they are not connected to the ground or another wire, current never flows through their bodies. Electric current stays in the wire. As long as a bird's body is only in contact with a line at the same voltage, it is safe. However, if a bird extends its wing and touches another wire at a different voltage, it would get a shock. Also, if a bird's wing touched a tree while it was sitting on a power line, it would get a shock.

Thinking Critically Why would a bird get shocked if its wing touched another wire at a different voltage?

6-8 What is an electric motor?

Objective

Explain how an electric motor works.

Key Term

electric motor: device that changes electrical energy into mechanical energy

Reversing Magnetic Poles By changing the direction of an electric current running through an electromagnet, you can reverse the poles of the electromagnet. Suppose that you connect an electromagnet to a dry cell. As the current flows, one end of the electromagnet becomes a north pole. The other end becomes a south pole. You can use a compass to identify the poles. What will happen if you switch the connections to the dry cell? The direction of the current will be reversed. The magnetic poles of the electromagnet will also be reversed. The compass will indicate the new magnetic poles.

1 PREDICT: What will happen to the poles of an electromagnet if you change the direction of the electric current?

Electric Motor The ability to reverse the poles of an electromagnet explains how an electric motor works. An **electric motor** changes electrical energy into mechanical energy. You learned earlier that mechanical energy is the energy of motion.

An electric motor is a device made up of an electromagnet, a permanent magnet, and a source of alternating current. The electromagnet is free to rotate. It is attached to a source of alternating current. When current flows, the poles of the electromagnet are attracted to the opposite poles of the permanent magnet. As the direction of the current changes, the poles of the electromagnet are reversed. The poles of the electromagnet are now repelled by the like poles of the permanent magnet. This repulsion causes the electromagnet to rotate.

Attraction and repulsion between the electromagnet and the permanent magnet cause the electromagnet to rotate. As long as alternating current is supplied, the electromagnet rotates continuously. The mechanical energy of the rotating magnet enables the motor to do work. The rotating magnet can be connected to a device such as a fan or a wheel.

2 INFER: How does an electric motor change electrical energy into mechanical energy?

Reversing Direct Current If direct current were used to run an electric motor, the electromagnet would not rotate continuously. Because the direction of the current would not change, the poles of the electromagnet would not reverse.

However, direct current can be used to run an electric motor if a special switch is used. This switch, called a commutator, reverses the direction of the current. Constantly reversing the direction of the current will cause the electromagnet to rotate.

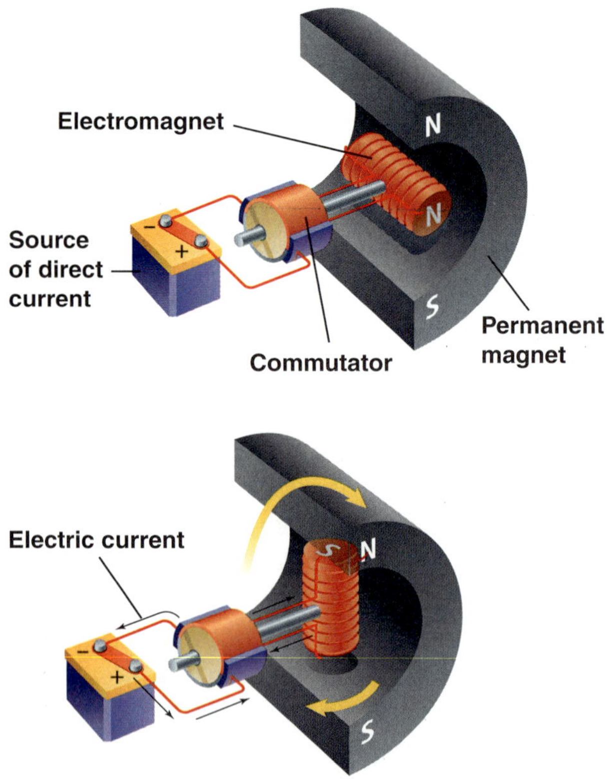

▲ **Figure 6-25** An electric motor at rest (top). When current flows (bottom), attraction and repulsion between the permanent magnet and the electromagnet cause the electromagnet to rotate. A commutator reverses the direct current to keep the electromagnet rotating.

3 INFER: What is the purpose of a commutator?

Uses of Electric Motors Many electrical appliances have electric motors. Electric fans, food processors, and refrigerators all use electric motors. There are many kinds of electric motors that are designed for specific uses. All electric motors have one thing in common. They all use the force of magnetism to change electrical energy into useful mechanical energy.

▲ **Figure 6-26** Electric toys use electric motors to run.

4 **EXPLAIN:** Why does an electrical appliance need a motor?

CHECKING CONCEPTS

1. The poles of an electromagnet are reversed when the __________ of an electric current is changed.
2. An electric motor changes electrical energy into __________ energy.
3. An electric motor contains an electromagnet and a __________ magnet.
4. The __________ in an electric motor is free to rotate.

THINKING CRITICALLY

5. **EXPLAIN:** Explain how an electric motor uses magnetism.

DESIGNING AN EXPERIMENT

Design an experiment to solve the following problem. Include a hypothesis, variables, a procedure, and a type of data to study.

PROBLEM: How can you design a toy or tool that uses an electric motor to run?

People in Science

AUTOMOBILE MECHANIC

▲ **Figure 6-27** An automobile mechanic maintains and repairs different parts of a car.

An automobile engine is a complex machine. It is made up of a motor, a battery, a generator, and many other electrical parts. Many jobs for automobile mechanics are available in automobile service stations, government installations, or in the service departments of automobile dealerships.

Automobile mechanics are trained to inspect, maintain, and repair the mechanical and electrical parts in different types of automobiles. They install and replace parts, adjust brakes, and use computers to check different parts of a car. Beginning mechanics get on-the-job training. Some receive classroom instruction together with hands-on practice. They may also work with an experienced mechanic. Specialty mechanics may work on only one part of a car's engine.

To be an automobile mechanic, you should graduate from high school and then complete a formal training program in automotive repair. Good reading, mathematics, and computer skills are usually required.

Thinking Critically Why do you think an automobile mechanic should have computer skills?

LAB ACTIVITY
Making an Electric Motor

Materials

- Safety goggles, apron
- D-cell battery
- Magnet wire, 1 meter
- Small film canister
- Fine sandpaper
- 2 large metal paper clips
- Ceramic magnet
- Wire cutters
- Needle-nose pliers
- Duct tape
- Modeling clay
- Pencil and paper

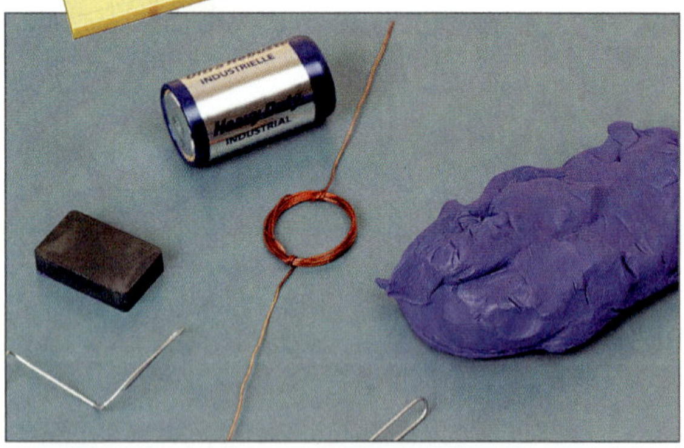

▲ **STEP 3** Wrap the coil to keep it from unwinding.

▲ **STEP 4** Sand off insulation from half of the wire.

BACKGROUND

Magnetism and electricity are closely related. Many devices you use every day, such as radios, CD players, and vacuum cleaners depend on this relationship. How can this relationship be demonstrated?

PURPOSE

You will construct a simple electric motor.

PROCEDURE

1. Copy the chart shown in Figure 6-28. Put on goggles and an apron.
2. Starting about 8 cm from one end of a piece of magnet wire, wrap the wire around a film canister seven times. Cut the long end of the wire, leaving an 8-cm end free.
3. Slide the coiled wire off the film canister and wrap the two free ends around the coil two or three times to keep the coil from unwinding.
4. Use fine sandpaper to remove all of the insulation from about 3 cm of one end of the wire. Then place the other end of the wire flat on your work surface and sand off the insulation from *half* of the wire for a length of about 3 cm.
5. Using needle-nose pliers, bend two large paper clips to make supports for the coil. Use duct tape to attach the paper clips to the ends of a D-cell battery.
6. Place a ceramic magnet on the side of the battery. then place the battery-magnet setup on a piece of modeling clay.

▲ **STEP 5** Bend two large paper clips.

▲ **STEP 8** Turn the coil slowly with your hand.

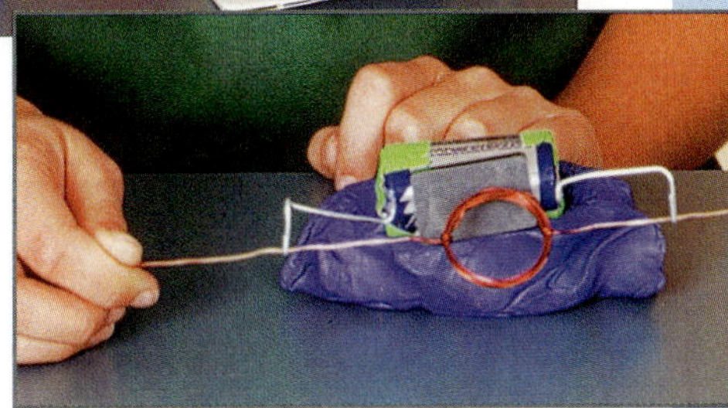

7. Place the coil in the cradle formed by the ends of the paper clip.
8. Turn the coil slowly with your hand. Record your observations.
9. Give the coil a gentle push to spin it. Record how many times the coil spins.

Making an Electric Motor

Experiment	Observations
Turning the coil by hand	
Spinning the coil	

▲ **Figure 6-28** Copy this chart and use it to record your observations.

CONCLUSIONS

1. **INFER:** What is the purpose of the battery and the magnet?
2. **OBSERVE:** What did you feel and observe as you moved the coil by hand?
3. **INFER:** What causes the coil to keep spinning after your initial push?
4. **ANALYZE:** What would happen if you were to increase the number of turns of wire?
5. **HYPOTHESIZE:** Is a circle the best shape? Do you think that an oval or square would be better? Give reasons for your answers.

6-9 What is an electric generator?

Objective

Explain how an electric generator works.

Key Term

electric generator: device that changes mechanical energy into electrical energy

Induced Current A current can be induced in a loop of wire by spinning the loop inside a magnetic field. Remember that an electric current is induced when a wire cuts across magnetic lines of force. Figure 6-29 shows a loop of wire in a magnetic field. The loop is spinning clockwise. As the wire moves across the magnetic field, it cuts the magnetic lines of force. As a result, current flows through the wire. When the wire moves down through the magnetic field, the current flows in one direction. When the wire moves up, the current flows in the opposite direction. The change in current direction produces an alternating current in the wire.

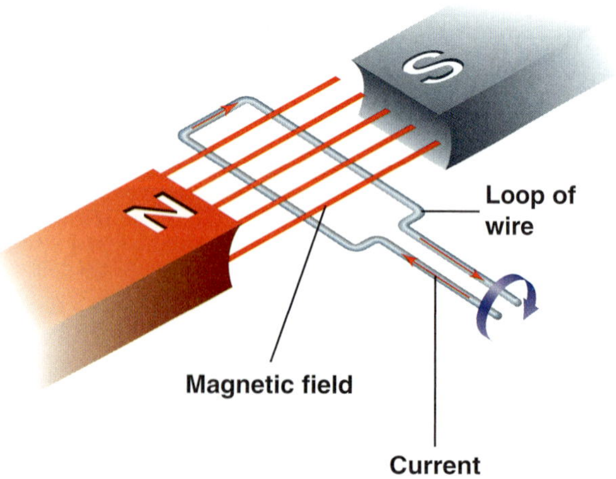

▲ **Figure 6-29** As the loop of wire spins inside the magnetic field of the magnets, alternating current is produced in the wire.

1 IDENTIFY: What kind of current is induced when a loop of wire spins in a magnetic field?

Electric Generators An **electric generator** is a device that changes mechanical energy into electrical energy. It is the opposite of an electric motor, which changes electrical energy to mechanical energy. An electric generator is made up of an insulated coil of wire and a magnet. The coil of wire is attached to a power source that causes the coil to spin. Spinning the coil of wire in the magnetic field of the magnet exposes the wire to a changing magnetic field. When this happens, electric current is produced in the coil of wire. The mechanical energy used to spin the coil of wire is changed into electrical energy. The ends of the coil of wire can be connected to a load, such as a light bulb. When electricity runs through the wires, the bulb will light.

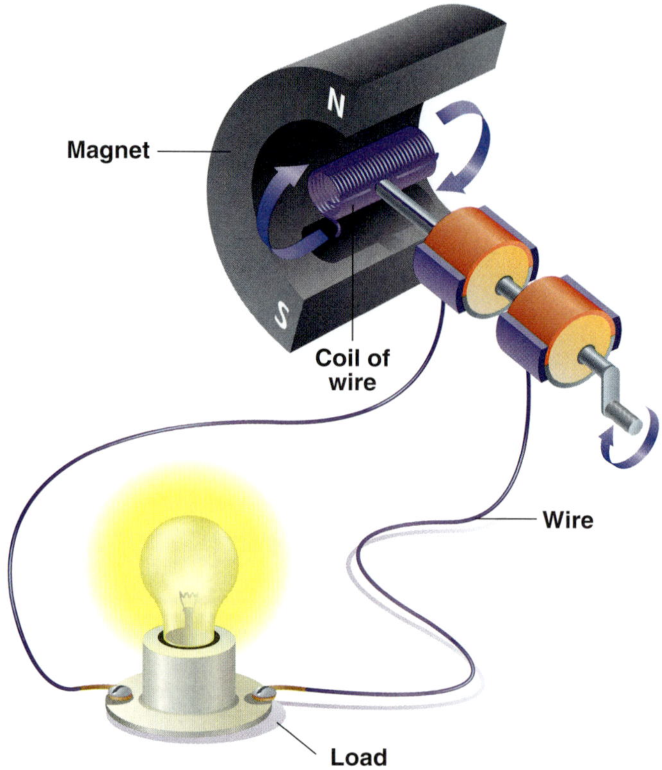

▲ **Figure 6-30** Turning the handle of this generator will cause the coil of wire to spin in the magnetic field of the magnet. An alternating electric current will be produced in the wires. As the current flows through the wires, the light bulb will light up.

2 DEFINE: What is an electric generator?

Uses of Generators Most of the electricity you use every day comes from generators. Power plants use large generators to supply electricity to homes, offices, schools, and other buildings. These large generators use very strong magnets and many coils of wire to produce a great amount of energy. The mechanical energy for these generators is supplied by turbines (TUR-bihnz). A turbine is a large wheel that is turned by steam or moving water. The water used to turn a turbine may come from a dam or river. Steam is produced by burning fuels, such as coal or oil, or from nuclear energy.

3 **NAME:** What are two fuels that are burned to make steam for turbines in generators?

CHECKING CONCEPTS

1. What happens when a loop of wire cuts magnetic lines of force?
2. What kind of electric current is induced when a loop of wire turns in a magnetic field?
3. What is a generator?
4. What supplies the mechanical energy in a generator?
5. What is a turbine?
6. Name two sources of energy that are used to make steam for turbines.

THINKING CRITICALLY

7. **CONTRAST:** Explain how a generator is the opposite of an electric motor.
8. **INFER:** What would happen if the coil of wire in a generator stopped spinning?
9. **HYPOTHESIZE:** The mechanical energy used to spin the coil of wire in a generator comes from a turbine. Think of ways in which you could supply the mechanical energy to run a turbine.

Web InfoSearch

Geothermal Energy In some parts of the world, heat inside Earth produces steam and boiling water. This heat is called geothermal (jee-oh-THUR-muhl) energy.

SEARCH: Use the Internet to find out how geothermal power plants use steam and boiling water to spin turbines and generate electricity. Start your search at www.conceptsandchallenges.com. Some key search words are **geothermal energy** and **generators.**

Science and Technology

ELECTRIC POWER PLANTS

▲ **Figure 6-31** Moving water spins turbines in this hydroelectric power plant.

Electric power plants provide electricity for large numbers of people. Electric power plants use generators to produce electricity. Instead of a simple loop of wire spinning in the magnetic field of a magnet, these generators use many coils of wire and strong electromagnets. The spinning wires are connected to a turbine. The energy to spin the turbine comes from steam or moving water.

Some generators use the energy from waterfalls to spin the turbine. These generators are known as hydroelectric (hy-droh-ee-LEHK-trihk) plants. Hydroelectric plants are usually built near dams and reservoirs. Hydroelectric plants provide only a limited amount of power in the United States. The high cost to build them and the damage to the environment that they cause are some reasons why hydroelectric power plants are not a major source of electricity.

Thinking Critically Why are hydroelectric power plants usually built near dams and reservoirs?

Integrating Technology

THE Big IDEA

How does a maglev train work?

Imagine using the power of magnets to make a train float above its tracks. It sounds a lot like science fiction but it is real. Trains that use electromagnets are called magnetic levitation trains, maglevs for short.

Currently, maglevs are an experimental form of transportation. These trains use the properties of magnets to float above their tracks. In some maglevs, electromagnets generate magnetic forces that cause the train and the track to repel each other. These magnetic forces are similar to the forces that cause the like poles of two magnets to repel each other. In other types of maglevs, a force of attraction is generated by electromagnets that lift up the train. This force of attraction is similar to the attraction between the unlike poles of two magnets. In both types of maglevs, the trains run on a cushion of air a few centimeters above their tracks.

There seem to be many advantages to maglev trains compared to conventional trains. Maglevs do not have many moving parts, and could, therefore, require less maintenance. Because maglevs float above their tracks, there is no friction between the train and the track. Without friction, maglevs produce less noise than conventional trains. Maglevs are also expected to be much faster than conventional trains. They can travel up to 500 km/h.

In Germany, engineers have designed a maglev train that uses attractive magnetic forces to lift up the train. Read about the German maglev train, the Transrapid, that appears on these two pages. Find out how this type of train operates. Then, follow the Science Log to learn more about "the big idea." ✦

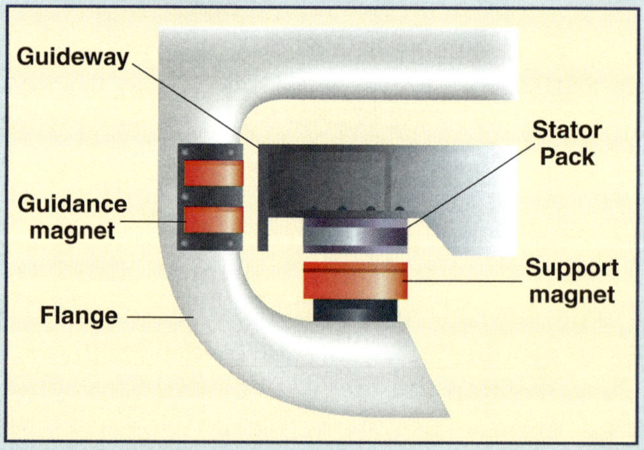

The Flanges

The bottom of the German Transrapid has extensions on both sides. These extensions, called flanges (FLAN-jez), wrap around a special track, called a guideway. The inside of a flange can be seen in the top photo. The diagram below it shows how magnets are used to lift the train. Magnetic attraction between the support magnet and the stator (STAT-uhr) pack, a highly magnetic material, causes the train to float about 10 mm above the guideway. Guidance magnets keep the train on the track.

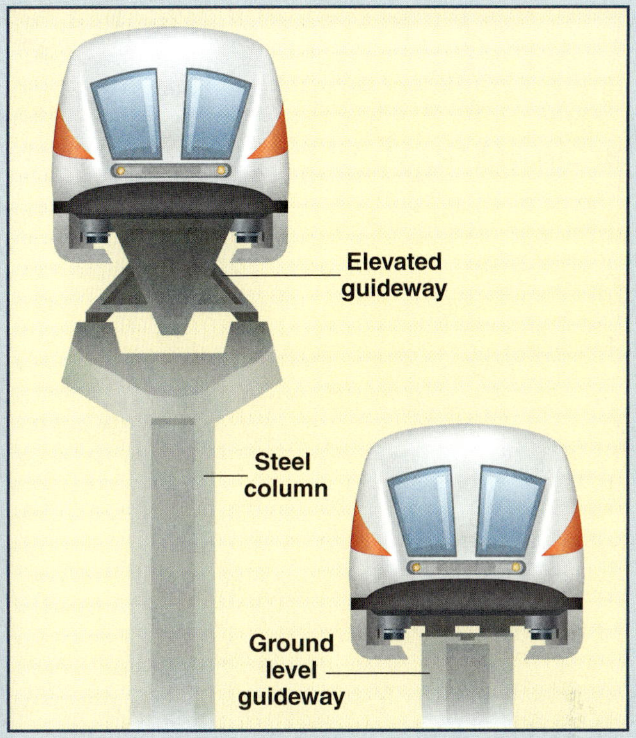

The Guideway

The guideway can be mounted on an elevated steel column (left illustration) or it can be mounted at ground level (right illustration).

The Transrapid

The German Transrapid maglev train can reach a maximum speed of about 500 km/h. The shape of its front portion helps it to travel fast by allowing it to move efficiently through air. It also cuts down on the amount of noise that the train makes.

WRITING ACTIVITY

Science Log

Maglev trains are expensive to build but they may provide a very fast way to travel between cities. Find out more about maglevs. In your science log, describe the advantages and disadvantages of this form of transportation. Compare maglev technology to cars, planes, and ordinary trains. Start your search at www.conceptsandchallenges.com.

Chapter 6 Challenges

Chapter Summary

Lesson 6-1

- **Magnetism** is a force of attraction or repulsion by magnetic materials.
- Like **poles** repel. Unlike poles attract.

Lesson 6-2

- A **magnetic field** is the area around a magnet where magnetic forces act.

Lesson 6-3

- The process by which a material is made into a magnet is called **magnetic induction.**
- A temporary magnet tends to lose its magnetism quickly.
- A permanent magnet tends to keep its magnetism.

Lesson 6-4

- Earth has a magnetic north pole and a magnetic south pole.
- The magnetic field around Earth is called the **magnetosphere.**

Lesson 6-5

- The relationship between electricity and magnetism is called **electromagnetism.**

Lesson 6-6

- An **electromagnet** is made by wrapping current-carrying coils of wire around a metal core.

Lesson 6-7

- A **transformer** uses alternating current in the primary coil to induce a current in the secondary coil.
- A step-up transformer increases the voltage of alternating current.
- A step-down transformer decreases the voltage of alternating current.

Lesson 6-8

- An **electric motor** is a device that changes electrical energy into mechanical energy.

Lesson 6-9

- An **electric generator** is a device that uses magnets to change mechanical energy into electrical energy.

Key Term Challenges

electric generator (p. 156)
electric motor (p. 152)
electromagnet (p. 148)
electromagnetic induction (p. 146)
electromagnetism (p. 146)
magnetic domain (p. 140)
magnetic field (p. 140)
magnetic induction (p. 142)
magnetic lines of force (p. 140)
magnetism (p. 138)
magnetosphere (p. 144)
pole (p. 138)
transformer (p. 150)

MATCHING Write the Key Term from above that best matches each description.

1. force of attraction or repulsion by magnetic materials
2. end of a magnet
3. temporary magnet formed by wrapping a current-carrying coil of wire around a metal core
4. process of making an object into a magnet

FILL IN Write the Key Term from above that best completes each statement.

5. Because of ___________, a current will flow in a wire moving across a magnetic field.
6. A ___________ converts mechanical energy into electrical energy.
7. The magnetic field around Earth is also called the ___________.
8. The area around a magnet is called the ___________.
9. The relationship between electricity and magnetism is called ___________.
10. An ___________ converts electrical energy into mechanical energy.

Content Challenges TEST PREP

MULTIPLE CHOICE Write the letter of the term or phrase that best completes each statement.

1. In a step-down transformer,
   a. the secondary coil has more turns.
   b. the primary coil has fewer turns.
   c. the secondary coil has fewer turns.
   d. the primary and secondary coils have the same number of turns.

2. Oersted discovered
   a. that an electric current flows through a compass.
   b. magnetic induction.
   c. the magnetosphere.
   d. that an electric current produces a magnetic field.

3. A temporary magnet
   a. keeps its magnetic properties for a long time.
   b. is hard to magnetize.
   c. loses its magnetic properties quickly.
   d. is a natural magnet.

4. When the north and south poles of two magnets are facing each other, they
   a. attract each other.
   b. repel each other.
   c. have no effect on each other.
   d. magnetize each other.

5. One way to increase the strength of an electromagnet is to
   a. increase the resistance in the wire.
   b. decrease the voltage in the wire.
   c. decrease the current in the wire.
   d. increase the current in the wire.

6. In order for a 9-volt appliance to use the 120 volts from a wall outlet, it needs
   a. a step-up transformer.
   b. a step-down transformer.
   c. a coil.
   d. an electric motor.

7. William Gilbert was one of the first scientists to study
   a. electric motors.
   b. generators.
   c. electromagnetism.
   d. magnetism.

8. Earth's magnetic north pole is located
   a. near the geographic North Pole.
   b. near the geographic South Pole.
   c. at the geographic North Pole.
   d. at the geographic South Pole.

9. An iron bar becoming magnetized is an example of
   a. magnetic induction.
   b. electromagnetic induction.
   c. a transformer.
   d. an electric motor.

10. Electromagnetic induction was studied by
    a. Tesla.
    b. Oersted.
    c. Faraday.
    d. the ancient Greeks.

TRUE/FALSE Write *true* if the statement is true. If the statement is false, change the underlined term to make the statement true.

11. In a step-up transformer, the <u>primary</u> coil has fewer turns of wire.

12. An electric <u>motor</u> converts electrical energy into mechanical energy.

13. To increase the strength of an electromagnet, <u>decrease</u> the number of turns of wire around the metal core.

14. Earth's magnetic south pole is the same as the <u>south</u> pole of a bar magnet.

Concept Challenges TEST PREP

WRITTEN RESPONSE Answer each of the following questions in complete sentences.

1. **COMPARE:** Explain the difference between magnetic induction and electromagnetic induction.
2. **CONTRAST:** Describe the energy changes that take place in an electric motor and in an electric generator.
3. **DESCRIBE:** How could you use a compass to identify the north and south poles of a magnet?
4. **ANALYZE:** Why does a spinning loop of wire in a magnetic field produce alternating current instead of direct current?
5. **HYPOTHESIZE:** In most transformers, some energy is lost between the primary coil and the secondary coil. What do you think might cause this loss of energy?

INTERPRETING VISUALS Use Figures 6-32 and 6-33 to answer each of the following questions.

6. In Diagram *A*, which coil has more turns of wire?
7. In Diagram *B*, which coil has more turns of wire?
8. Which diagram shows a step-up transformer? Which shows a step-down transformer?
9. Which kind of transformer would you use to change the voltage from 10 volts to 50 volts? Explain.
10. Which kind of transformer would you use to change the voltage from 50 volts to 10 volts? Explain.
11. Which kind of transformer must be used before the electricity from power lines can be used in your home? Explain.

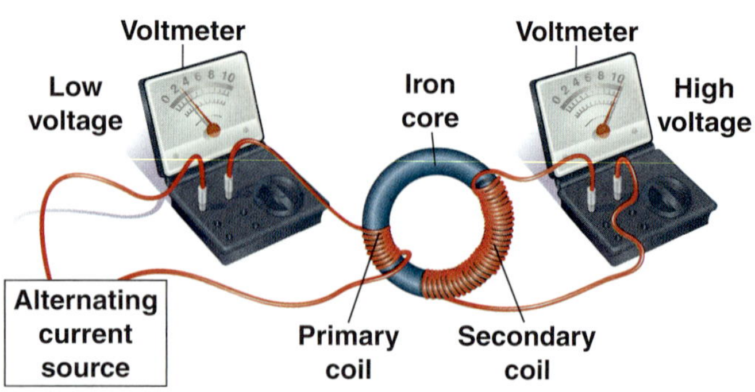

▲ **Figure 6-32** Diagram A

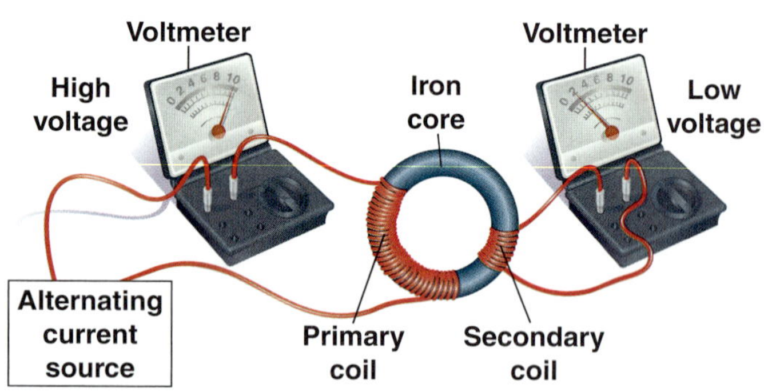

▲ **Figure 6-33** Diagram B

Appendix A Metric System

The Metric System and SI Units

The metric system is an international system of measurement based on units of ten. More than 90% of the nations of the world use the metric system. In the United States, both the English system and the metric system are used.

The *Système International*, or SI, has been used as the international measurement system since 1960. The SI is a modernized version of the metric system. Like the metric system, the SI is a decimal system based on units of ten. When you want to change from one unit in the metric system to another unit, you multiply or divide by a multiple of ten.

- When you change from a smaller unit to a larger unit, you divide.
- When you change from a larger unit to a smaller unit, you multiply.

METRIC UNITS		
LENGTH	SYMBOL	RELATIONSHIP
kilometer	km	1 km = 1,000 m
meter	m	1 m = 100 cm
centimeter	cm	1 cm = 10 mm
millimeter	mm	1 mm = 0.1 cm
AREA	SYMBOL	
square kilometer	km^2	1 km^2 = 1,000,000 m^2
square meter	m^2	1 m^2 = 1,000,000 mm^2
square centimeter	cm^2	1 cm^2 = 0.0001 m^2
square millimeter	mm^2	1 mm^2 = 0.000001 m^2
VOLUME	SYMBOL	
cubic meter	m^3	1 m^3 = 1,000,000 cm^3
cubic centimeter	cm^3	1 cm^3 = 0.000001 m^3
liter	L	1 L = 1,000 mL
milliliter	mL	1 mL = 0.001 L
MASS	SYMBOL	
metric ton	t	1 t = 1,000 kg
kilogram	kg	1 kg = 1,000 g
gram	g	1 g = 1,000 mg
centigram	cg	1 cg = 10 mg
milligram	mg	1 mg = 0.001 g
TEMPERATURE	SYMBOL	
Kelvin	K	
degree Celsius	°C	

▲ Figure 1

COMMON METRIC PREFIXES			
micro-	0.000001 or 1/1,000,000	deka-	10
milli-	0.001 or 1/1,000	hecto-	100
centi-	0.01 or 1/100	kilo-	1,000
deci-	0.1 or 1/10	mega-	1,000,000

▲ Figure 2

METRIC-STANDARD EQUIVALENTS	
SI to English	English to SI
LENGTH	
1 kilometer = 0.621 mile (mi)	1 mi = 1.61 km
1 meter = 1.094 yards (yd)	1 yd = 0.914 m
1 meter = 3.28 feet (ft)	1 ft = 0.305 m
1 centimeter = 0.394 inch (in.)	1 in. = 2.54 cm
1 millimeter = 0.039 inch	1 in. = 25.4 mm
AREA	
1 square kilometer = 0.3861 square mile	1 mi^2 = 2.590 km^2
1 square meter = 1.1960 square yards	1 yd^2 = 0.8361 m^2
1 square meter = 10.763 square feet	1 ft^2 = 0.0929 m^2
1 square centimeter = 0.155 square inch	1 $in.^2$ = 6.452 cm^2
VOLUME	
1 cubic meter = 1.3080 cubic yards	1 yd^3 = 0.7646 m^3
1 cubic meter = 35.315 cubic feet	1 ft^3 = 0.0283 m^3
1 cubic centimeter = 0.0610 cubic inch	1 $in.^3$ = 16.39 cm^3
1 liter = 0.2642 gallon (gal)	1 gal = 3.79 L
1 liter = 1.06 quarts (qt)	1 qt = 0.946 L
1 liter = 2.11 pints (pt)	1 pt = 0.47 L
1 milliliter = 0.034 fluid ounce (fl oz)	1 fl oz = 29.57 mL
MASS	
1 metric ton = 0.984 ton	1 ton = 1.016 t
1 kilogram = 2.205 pounds (lb)	1 lb = 0.4536 kg
1 gram = 0.0353 ounce (oz)	1 oz = 28.35 g
TEMPERATURE	
Celsius = 5/9(°F – 32)	Fahrenheit = 9/5°C + 32
0°C = 32°F (Freezing point of water)	72°F = 22°C (Room temperature)
100°C = 212°F (Boiling point of water)	98.6°F = 37°C (Human body temperature)
Kelvin = (°F + 459.67)/1.8	Fahrenheit = (K × 1.8) – 459.67

▲ Figure 3

Appendix B Science Terms

Analyzing Science Terms

You can often unlock the meaning of an unfamiliar science term by analyzing its word parts. Prefixes and suffixes, for example, each carry a meaning that comes from a word root. This word root usually comes from the Latin or Greek language. The following list of prefixes and suffixes provides clues to the meaning of many science terms.

WORD PART	MEANING	EXAMPLE
-ate	salt of an acid	nitrate
bar-, baro-	weight, pressure	barometer
bi-	two	binary
carbo-	containing carbon	carbonate
co-	with, together	coagulation
de-	remove from	decomposition
electro-	electricity	electrolyte
-graph	write	thermograph
hydro-	water, containing hydrogen	hydrometer, hydrocarbon
-ide	binary compound	sulfide
in-	not	insoluble
-logy	study of	cosmology
-lysis	decomposition	electrolysis
magneto-	magnetism	magnetosphere
-meter	measuring device	manometer
non-	not	nonmetal
photo-	light	photoelectric
poly-	many	polyatomic
re-	again, back	reflection
-sonic	sound	supersonic
-sphere	ball, globe	magnetosphere
sub-	under, beneath	subscript
super-	above, more than	supersonic
therm-, thermo-	heat	thermometer
trans-	across, beyond	transparent
ultra-	beyond	ultrasound
un-	not	unsaturated

▲ Figure 4

Appendix C Mathematics Review

Adding Integers

You can add integers with unlike signs on a number line.

Add ${}^{-}5 + {}^{+}7$

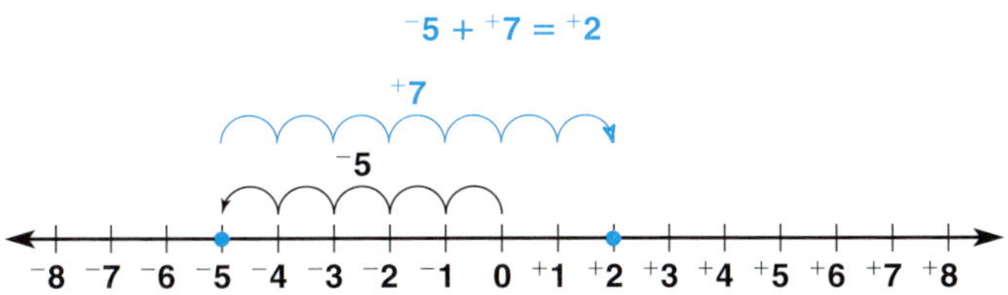

Subtracting Integers

To subtract an integer, add its opposite.

Subtract ${}^{-}6 - {}^{+}2$

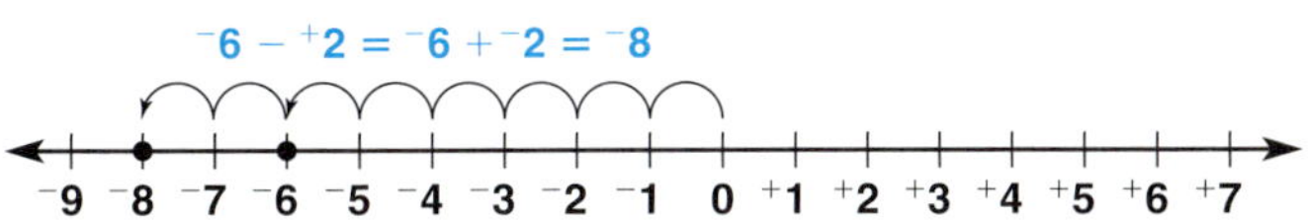

Multiplying Integers

When you multiply integers, you must decide if the answer is positive or negative.

If the signs of the integers are the same, the product is positive.

$${}^{+}5 \times {}^{+}4 = {}^{+}20$$

$${}^{-}5 \times {}^{-}4 = {}^{+}20$$

If the signs of the integers are different, the product is negative.

$${}^{+}5 \times {}^{-}4 = {}^{-}20$$

$${}^{-}5 \times {}^{+}4 = {}^{-}20$$

Dividing Integers

The rules for dividing integers are the same as the rules for multiplying integers.

If the signs of the integers are the same, the quotient is positive.

$${}^{-}36 \div {}^{-}9 = {}^{+}4$$

$${}^{+}36 \div {}^{+}9 = {}^{+}4$$

If the signs of the integers are different, the quotient is negative.

$${}^{-}36 \div {}^{+}9 = {}^{-}4$$

$${}^{+}36 \div {}^{-}9 = {}^{-}4$$

Solving an Equation

To solve an equation, find the value of the variable that makes the equation true.

Is $b = 3$ the solution to the equation?

$$4b = 12$$

Replace b with 3 in the equation.

$$4 \times 3 = 12$$

$$12 = 12$$

Yes, $b = 3$ is the solution to the equation.

Adding and Subtracting Decimals

When adding or subtracting decimals, always be sure to line up the decimal points correctly.

Add 3.4 km, 20.95 km, and 153.6 km.

$$\begin{array}{r} 3.4 \\ 20.95 \\ +\ 153.6 \\ \hline 177.95 \text{ km} \end{array}$$

Subtract 13.5 mL from 35.75 mL.

$$\begin{array}{r} 35.75 \\ -\ 13.5 \\ \hline 22.25 \text{ mL} \end{array}$$

Multiplying and Dividing Decimals

When multiplying or dividing decimals, it is not necessary to line up the decimal points.

Multiply 0.5 N by 11.25 m to find the amount of work done in joules.

$W = F \times d$

$W = 0.5\ N \times 11.25\ m$

$W = 5.625\ J$

Notice that the number of places to the right of the decimal point in the answer is equal to the sum of the places to the right of the decimal point in the numbers being multiplied.

Divide 4.05 m by 0.5 m to find the mechanical advantage of a lever.

MA = effort arm length/resistance arm length

MA = 4.05 m/0.5 m

MA = 8.1

When dividing a decimal by another decimal, you must first change the divisor to a whole number. For example, change 0.5 to 5 by moving the decimal point one place to the right. You must also change the dividend by moving the decimal point one place to the right. The result is $40.5 \div 5 = 8.1$.

Changing a Decimal to a Percent

To change a decimal to a percent, multiply the decimal by 100%.

Find the efficiency of a machine if the work output is 5 J and the work input is 10 J.

Efficiency = work output $\div$ work input $\times$ 100%

Efficiency = 5 J $\div$ 10 J $\times$ 100%

Efficiency = 0.5 $\times$ 100%

Efficiency = 50%

Notice that when you multiply 0.5 by 100%, the decimal point moves two places to the right.

Measuring Angles

Use a protractor to measure an angle. Place the center of the protractor's straight edge on the vertex. One ray must pass through 0°.

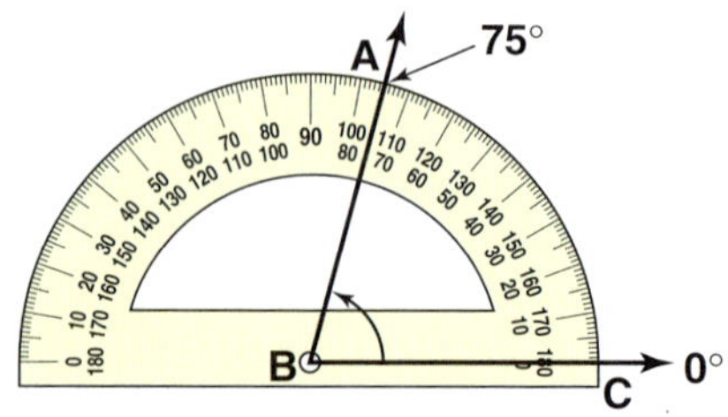

Angle ABC measures 75°.

Solving Word Problems

To solve distance problems, you can use $d = r \times t$ or $d = rt$.

The Smiths drove 220 miles at an average speed of 55 miles per hour. How long did the trip take?

PLAN

Substitute the values you know into the equation $d = r \times t$.

Then solve.

DO

$$220 = 55t$$

$$220 \div 55 = 55t \div 55$$

$$4 = t$$

SOLUTION

The trip took 4 hours.

Appendix D Mathematical Formulas

FORCE AND MOTION

Speed

Velocity is speed and direction.

1. Speed = distance ÷ time

   or

   $v = \frac{d}{t}$

   or

   $v = d \div t$

2. Change in velocity = final velocity − initial velocity
3. Acceleration = change in velocity ÷ time

Distance

Distance traveled is measured in miles or kilometers per hour (mph or km/h).

Distance = rate × time

or

$d = r \times t$

or

$d = rt$

Momentum

Momentum = mass × velocity

Force

Force = mass × acceleration

Force = pressure × area

Pressure

Pressure is expressed in pascals.

$1 \text{ Pa} = 1 \frac{N}{m^2}$

Pressure = force ÷ area

or

$P = \frac{F}{a}$

or

$P = F \div a$

ENERGY

Energy

Energy = mass × speed of light^2

or

$E = m \times c^2$

or

$E = mc^2$

Ohm's Law

Current is measured in amperes, or amps.

Current = voltage ÷ resistance

or

$I = \frac{V}{R}$

or

$I = V \div R$

Gravitational Potential Energy

PE = weight × height

PE (N-m) = N × m

Speed of a Wave

Velocity = frequency × wavelength

$v = f \times D$

WORK

Mechanical Advantage

1. Mechanical Advantage = resistance force ÷ effort force

   or

   $MA = \frac{\text{resistance force}}{\text{effort force}} = \frac{F_r}{F_e}$

   or

   $MA = F_r \div F_e$

2. MA (of an inclined plane) = length × height

   $MA = l \times h$

   $MA = lh$

Appendix D Mathematical Formulas (continued)

Efficiency

Efficiency is expressed as a percentage.

$$\text{Efficiency} = \text{Work output} \div \text{Work input} \times 100\%$$

or

$$E = \frac{W_{out}}{W_{in}}$$

or

$$E = W_{out} \div W_{in} \times 100\%$$

Work

Work is measured in joules (J). One joule equals one newton-meter (N-m).

$$W = \text{force} \times \text{distance}$$

or

$$W = F \times d$$

Power

Power is measured in watts (W).

One watt is equal to one joule per second (J/s).

$$\text{Power} = \text{Work} \div \text{time}$$

or

$$P = \frac{W}{t}$$

or

$$P = W \div t$$

Glossary

Pronunciation and syllabication have been derived from *Webster's New World Dictionary*, Second College Edition, Revised School Printing (Prentice Hall, 1985). Syllables printed in capital letters are given primary stress. (Numbers in parentheses indicate the page number, or page numbers, on which the term is defined.)

PRONUNCIATION KEY					
Symbol	**Example**	**Respelling**	**Symbol**	**Example**	**Respelling**
a	transverse	(trans-VURS)	oh	coagulation	(koh-ag-yoo-LAY-shuhn)
ah	velocity	(vuh-LAHS-uh-tee)	oo	amplitude	(AM-pluh-tood)
aw	trough	(TRAWF)	oi	colloid	(KAHL-oid)
ay	radiation	(ray-dee-AY-shuhn)	s	solute	(SAHL-yoot)
eh	convection	(kuhn-VEHK-shuhn)	sh	suspension	(suh-SPEHN-shuhn)
ee	decomposition	(dee-kahm-puh-ZIH-shuhn)	u	fulcrum	(FOOL-kruhm)
f	coefficient	(koh-uh-FIHSH-uhnt)	uh	barometer	(buh-RAHM-uht-uhr)
ih	specialization	(spehsh-uhl-ih-ZAY-shuhn)	y, eye	binary, ion	(BY-nuh-ree), (EYE-uhn)
j	homogenization	(huh-mahj-uh-nih-ZAY-shuhn)	yoo	insoluble	(ihn-SAHL-yoo-buhl)
k	calorie	(KAL-uh-ree)	z	ionization	(eye-uh-nih-ZAY-shuhn)

A

alternating current: current in which electrons constantly change direction at a regular rate (p. 120)

ampere: unit used to measure electric current (p. 126)

amplitude (AM-pluh-tood)**:** height of a transverse wave (p. 46)

B

battery: a series of electrochemical cells that are connected to each other (p. 116)

boiling point: temperature at which a liquid changes to a gas (p. 22)

C

calorie (KAL-uh-ree)**:** unit of heat; amount of heat needed to raise the temperature of 1 g of water 1°C (p. 18)

chlorophyll (KLAWR-uh-fihl)**:** substance in plants that absorbs the Sun's light and gives plants their green color (p. 92)

chloroplast: part of green plant cells where photosynthesis takes place (p. 92)

circuit breaker: switch that opens a circuit if too much current is flowing (p. 130)

cochlea (KAHK-lee-uh)**:** organ that changes sound vibrations into nerve signals (p. 78)

communication: sharing information (p. 8)

compression (kuhm-PREHSH-uhn)**:** part of a medium where the particles are close together (p. 44)

concave lens: lens that curves inward (p. 94)

conduction (kuhn-DUK-shuhn)**:** process of heat transfer in solids (p. 24)

conductor: material that conducts heat easily (p. 24); material through which electric charge can flow easily (p. 118)

constant: something that does not change (p. 11)

controlled experiment: experiment in which all the conditions except one are kept constant (p. 11)

convection (kuhn-VEHK-shuhn)**:** transfer of heat in gases and liquids (p. 26)

convection current: movement of gases or liquids caused by differences in density (p. 26)

convex lens: lens that curves outward (p. 94)

crest: high point of a transverse wave (p. 44)

D

data: information you collect when you observe something (p. 3)

decibel: unit used to measure the intensity or loudness of a sound (p. 68)

diffuse reflection: reflection that forms a fuzzy image (p. 100)

direct current: current in which electrons always flow in the same direction (p. 120)

dispersion (di-SPUR-shuhn)**:** separation of white light into its component colors (p. 102)

Doppler effect: apparent change in the frequency of waves (p. 76)

E

ear: sense organ that detects sound (p. 78)

eardrum: thin sheet of tissue that vibrates when sound waves strike it (p. 78)

echo: reflected sound waves (p. 64)

electric circuit: path that an electric current follows (p. 122)

electric current: flow of electric charge through a conductor (p. 120)

electric generator: device that changes mechanical energy into electrical energy (p. 156)

electric motor: device that changes electrical energy into mechanical energy (p. 152)

electrochemical cell: device that changes chemical energy into electrical energy (p. 116)

electrode: negative or positive pole of an electrochemical cell (p. 116)

electrolyte (ee-LEHK-troh-lyt)**:** substance that forms ions when melted or dissolved in water (p. 116)

electromagnet: temporary magnet made by wrapping a current-carrying wire around a metal core (p. 148)

electromagnetic induction: process by which an electric current is produced by moving a wire in a magnetic field (p. 146)

electromagnetic (ee-lehk-troh-mag-NEHT-ihk) **spectrum:** range of electromagnetic waves (p. 84)

electromagnetic wave: wave that transfers energy through empty space (p. 42)

electromagnetism: relationship between electricity and magnetism (p. 146)

electron: atomic particle with a negative electrical charge (p. 112)

evaporation (ee-vap-uh-RAY-shuhn)**:** change from a liquid to a gas at the surface of the liquid (p. 22)

eye: sense organ that detects light (p. 96)

F

fiber optics: use of optical fibers (p. 104)

freezing point: temperature at which a liquid changes to a solid (p. 20)

frequency (FREE-kwuhn-see)**:** number of complete waves passing a point in a given time (p. 46)

fundamental (fun-duh-MEHN-tuhl) **tone:** lowest pitched sound produced when an object vibrates (p. 74)

fuse: thin piece of metal that melts and breaks a circuit if too much current is flowing (p. 130)

G

gram: basic unit of mass (p. 4)

H

hearing: one of the five human senses (p. 78)

heat: energy of particles moving from warmer regions to cooler regions (p. 16)

hertz (HURTS)**:** unit used to measure the frequency of a wave (p. 46)

hypothesis: suggested answer to a question or problem (p. 10)

I

illuminated (ih-LOO-muh-nayt-uhd) **object:** object that reflects light (p. 90)

image: picture formed by the eye (p. 96)

incident wave: wave that strikes a barrier (p. 50)

infrasonic (ihn-fruh-SAHN-ihk)**:** sound below 20 Hz frequency (p. 70)

insulator: material that does not conduct heat easily (p. 24); material through which electric charge does not flow easily (p. 118)

intensity: amount of energy in a sound wave (p. 68)

L

laser: device that produces a powerful beam of light (p. 104)

lens: transparent material that bends light (p. 94)

liter: basic unit of liquid volume (p. 4)

light: form of electromagnetic energy made up of streams of photons (p. 86)

longitudinal (lahn-juh-TOOD-uhn-uhl) **wave:** wave in which the particles of the medium move back and forth in the direction of the wave motion (pp. 44, 64)

luminous (LOO-muh-nuhs) **object:** object that gives off its own light (p. 90)

M

magnetic domain: groups of atoms whose magnetic poles are all lined up in the same direction (p. 140)

magnetic field: area around a magnet where magnetic forces act (p. 140)

magnetic induction: process by which a material can be made into a magnet (p. 142)

magnetic lines of force: lines that show the shape of a magnetic field (p. 140)

magnetism: force of attraction or repulsion by magnetic materials (p. 138)

magnetosphere (mag-NEET-oh-sfeer)**:** region of Earth's magnetic field (p. 144)

mass: amount of matter in something (p. 4)

mechanical wave: wave that transfers energy through matter (p. 42)

medium: material through which mechanical waves can travel (pp. 42, 64)

melting point: temperature at which a solid changes to a liquid (p. 20)

meniscus: curve at the surface of a liquid in a thin tube (p. 4)

meter: basic unit of length or distance (p. 4)

model: tool scientists use to represent an object or a process (p. 3)

music: a pleasing combination of sounds (p. 74)

N

neutron: atomic particle with neither a negative nor a positive electric charge (p. 112)

noise: unpleasant combination of sounds with irregular patterns of vibration (p. 74)

normal: line at right angles to a barrier (p. 50)

O

ohm: unit used to measure resistance (p. 126)

Ohm's law: current in a wire is equal to the voltage divided by the resistance (p. 128)

opaque (oh-PAYK)**:** material that blocks light (p. 90)

overtones: sounds that have a higher pitch than the fundamental tone (p. 74)

P

parallel circuit: circuit in which an electric current can follow more than one path (p. 124)

photon (FOH-tahn)**:** tiny bundle of energy (p. 86)

photosynthesis (foht-oh-SIHN-thuh-sihs)**:** process by which plants use energy from the Sun to make food (p. 92)

pitch: how high or low a sound is (p. 70)

plane mirror: smooth surface that reflects light and forms images (p. 100)

pole: end of a magnet (p. 138)

prism (PRIZ-uhm)**:** triangular piece of transparent glass that separates white light into a band of colors (p. 102)

proton: atomic particle with a positive electric charge (p. 112)

R

radiation (ray-dee-AY-shuhn)**:** transfer of energy through space (p. 30)

rarefaction (rer-uh-FAK-shuhn)**:** part of a medium where the particles are far apart (p. 44)

ray: straight line that shows the direction of light (p. 86)

real image: image that can be projected onto a screen (p. 94)

reflected wave: wave that bounces off a barrier (p. 50)

reflection: bouncing back of a wave after striking a barrier (p. 50)

refraction: bending of a wave as it moves from one medium to another (p. 52)

regular reflection: reflection that forms a clear image (p. 100)

resistance: opposition to the flow of electric charges (p. 126)

S

series circuit: circuit in which electric current follows only one path (p. 122)

sight: one of the five human senses (p. 96)

simulation: computer model that usually shows a process (p. 3)

sonic boom: loud noise caused by breaking the sound barrier (p. 66)

sound: form of energy that travels as waves (p. 60)

specific heat: amount of heat needed to raise the temperature of 1 g of a substance 1°C (p. 32)

speed: distance a wave travels in one unit of time (p. 46)

static electricity: buildup of electric charges in an object (p. 112)

supersonic: faster than the speed of sound for that medium (p. 66)

T

temperature: measurement of the amount of heat energy something contains (pp. 4, 18)

theory: set of hypotheses that have been supported by testing over and over again (p. 10)

thermal (THUR-muhl) **expansion:** expansion of matter caused by heating (p. 34)

timbre (TAM-buhr)**:** sound quality (p. 74)

total internal reflection: occurs when light is repeatedly reflected within a given material (p. 104)

transformer: device in which alternating current in one coil of wire induces a current in a second coil of wire (p. 150)

translucent (trans-LOO-suhnt)**:** material that transmits some light (p. 90)

transparent (trans-PEHR-uhnt)**:** material that transmits light easily (p. 90)

transverse (trans-VURS) **wave:** wave in which the particles of the medium move up and down at right angles to the direction of the wave motion (pp. 44, 88)

trough (TRAWF)**:** low point of a transverse wave (p. 44)

U

ultrasonic (uhl-truh-SAHN-ihk)**:** sound above 20,000 Hz frequency (p. 70)

unit: amount used to measure something (p. 4)

V

vacuum: region where no matter exists (p. 30)

variable: anything that can affect the outcome of an experiment (p. 11)

vibration: rapid back-and-forth movement (p. 60)

virtual image: image that cannot be projected onto a screen (p. 94)

visible spectrum: seven colors that make up white light (p. 84)

volt: unit used to measure voltage (p. 126)

voltage: energy available to move charges through a circuit (p. 126)

volume: amount of space an object takes up (p. 4)

W

wave: disturbance that transfers energy from place to place (p. 42)

wavelength: distance between two neighboring crests or troughs (p. 46)

Index

Photo Credits

Photography Credits: All photographs are by the Pearson Learning Group (PLG), John Serafin for PLG, and David Mager for PLG, except as noted below.

Cover: *bkgd.* Tek Image/Science Photo Library/Photo Researchers, Inc.; *inset* Warren Bolster/Getty Images, Inc.

Table of Contents: iv: b S. Dalton, OSF/Animals Animals/Earth Scenes; v: t Philippe Plail/Photo Researchers, Inc.; v: b Unknown/Con Edison of New York

Frontmatter: P001 bl CityNet Telecom; P001 bm John Sohlden/Visuals Unlimited, Inc.; P001 tr Gregg Otto/Visuals Unlimited, Inc.; P002 l Bernd Wittich/Visuals Unlimited, Inc.; P001 Stock Trek/PhotoDisc, Inc.; P002 m Charles O'Rear/Corbis; P003 Stock Trek/PhotoDisc, Inc.; P005 r Stock Trek/PhotoDisc, Inc.; P007 r Stock Trek/PhotoDisc, Inc.; P009 r Stock Trek/PhotoDisc, Inc.; P009 mr Larry Mulvehill/Photo Researchers, Inc.; P009 t Jay Freis/Image Bank; P009 tr Ed Young/Corbis; P011 r Stock Trek/PhotoDisc, Inc.; P013 r Stock Trek/PhotoDisc, Inc.

Chapter 1: P15 Michael Fairchild/Peter Arnold, Inc.; P17 l Science VU/Visuals Unlimited, Inc.; P17 r E. R. Degginger/Color Pic, Inc.; P18 Mark E. Gibson/Visuals Unlimited, Inc.; P19 The Granger Collection, New York; P21 b Jay Freis/Image Bank; P21 t Jerome Wexler/Visuals Unlimited, Inc.; P24 b Phil Degginger/Color Pic, Inc.; P25 b Phil Degginger/Color Pic, Inc.; P25 t Phil Degginger/Color Pic, Inc.; P26 Johnny Johnson/DRK Photo; P30 Phil Degginger/Color Pic, Inc.; P31 t Pat Armstrong/Visuals Unlimited, Inc.; P34 l Richard Choy/Peter Arnold, Inc.; P34 r E. R. Degginger/Color Pic, Inc.; P36 b Rod Planck/Photo Researchers, Inc.; P36 t Fritz Polking/Visuals Unlimited, Inc.; P37 bl A. & M. Shah/Animals Animals/Earth Scenes; P37 br Norbert Wu/Peter Arnold, Inc.; P37 t Dan Guravich/Photo Researchers, Inc.; P38 Michael Fairchild/Peter Arnold, Inc.; P39 Michael Fairchild/Peter Arnold, Inc.; P40 Michael Fairchild/Peter Arnold, Inc.

Chapter 2: P41 George Bernard/Animals Animals/Earth Scenes; P42 l Tony Arruza/Corbis; P42 r Nina Leen/Time Life Syndication; P47 Hulton-Deutsch Collection/Corbis; P50 Phil Degginger/Color Pic, Inc.; P54 inset Corbis; P54–55 bkgd Burstein Collection/Corbis; P55 inset l Ken M. Johns/Photo Researchers, Inc.; P55 inset m Courtesy Pacific Tsunami Museum; P55 inset r State of Hawaii, Department of Defense, Civil Defense Division; P56 George Bernard/Animals Animals/Earth Scenes; P57 George Bernard/Animals Animals/Earth Scenes; P58 George Bernard/Animals Animals/Earth Scenes

Chapter 3: P59 E. R. Degginger/Color Pic, Inc.; P64 NASA/Omni-Photo Communications; P65 Robert Marien/Index Stock Imagery; P66 l Michael Fredericks, Jr./Animals Animals/Earth Scenes; P66 r Robert A. Hoover/NASA/Dryden Flight Center; P67 b Richard Neville/Check Six; P67 t Eric Risberg/AP Photo; P69 Robert Brenner/PhotoEdit, Inc.; P70 b Jack Parsons/Omni-Photo Communications; P70 t E. R. Degginger/Color Pic, Inc.; P71 Mark Newman/Photo Researchers, Inc.; P72 b Francois Gohier/Photo Researchers, Inc.; P72 t E. R. Degginger/Animals Animals/Earth Scenes; P73 bl William Lampas/Omni-Photo Communications; P73 br Bruna Stude/Omni-Photo Communications; P73 t S. Dalton, OSF/Animals Animals/Earth Scenes; P77 National Optical Astronomy Observatories/Colored by Science Photo Library/Photo Researchers, Inc.; P79 AT&T Archives; P80 E. R. Degginger/Color Pic, Inc.; P81 E. R. Degginger/Color Pic, Inc.; P82 E. R. Degginger/Color Pic, Inc.

Chapter 4: P83 Tek Image/Science Photo Library/Photo Researchers, Inc.; P89 Janis Christie/PhotoDisc, Inc.; P93 David Julian/Phototake; P95 Robert Garvey/Corbis; P97 Alexander Tsiaras/Science Course/Photo Researchers, Inc.; P100 l Jim Steinberg/Photo Researchers, Inc.; P100 r Jeremy Woodhouse/PhotoDisc, Inc.; P103 David M. Schleser/Nature's Images/Photo Researchers, Inc.; P104 Michael Rizza/Stock, Boston Inc.; P105 Philippe Plail/Photo Researchers, Inc.; P108 Tek Image/Science Photo Library/Photo Researchers, Inc.; P109 Tek Image/Science Photo Library/Photo Researchers, Inc.; P110 Tek Image/Science Photo Library/Photo Researchers, Inc.

Chapter 5: P111 Fotopic/Omni-Photo Communications; P117 Ecoscene/Corbis; P119 Bettmann/Corbis; P121 b Fritz Prenzel/Animals Animals/Earth Scenes; P121 t Richard T. Nowitz/Photo Researchers, Inc.; P123 Science VU/Honeywell/Visuals Unlimited, Inc.; P125 Rick Poley/Visuals Unlimited, Inc.; P127 b Phil Degginger/Color Pic, Inc.; P127 t E. R. Degginger/Color Pic, Inc.; P128 Richard Megna/Fundamental Photographs; P129 Grantpix/Photo Researchers, Inc.; P130 l Phil Degginger/Color Pic, Inc.; P130 m E. R. Degginger/Color Pic, Inc.; P130 r Phil Degginger /Color Pic, Inc.; P131 GJLP/CNRI/Phototake; P132 inset b Bettmann/Corbis; P132 inset t Michael Freeman/Corbis; P132-133 Getty Images/Hulton Archive; P133 inset l Unknown/Con Edison of New York; P133 inset r Hulton-Deutsch Collection/Corbis; P134 Fotopic/Omni-Photo Communications; P135 Fotopic/Omni-Photo Communications; P136 Fotopic/Omni-Photo Communications

Chapter 6: P137 E. R. Degginger/Color Pic, Inc.; P138 Michael Newman/Photo Edit, Inc.; P139 Tony Craddock/Science Photo Library/Photo Researchers, Inc.; P140 b Biodisc/Visuals Unlimited, Inc.; P140 t Biodisc/Visuals Unlimited, Inc.; P141 Biodisc/Visuals Unlimited, Inc.; P142 E. R. Degginger/Color Pic, Inc.; P145 b Unknown/Hamblin Publishing Co., Inc.; P145 t NASA/Science Photo Library/Photo Researchers, Inc.; P147 National Portrait Gallery, Washington DC/Art Resource; P149 Spencer Grant/Photoedit, Inc.; P151 Scott T. Smith/ Corbis; P153 b Tony Freeman/Photo Edit/Picturequest; P153 t Mark E. Gibson/Visuals Unlimited, Inc.; P157 Jeff Greenberg/Omni-Photo Communications; P158 AP/Wide World Photos; P159 b Thomas Raupach/Woodfin Camp & Associates; P159 t Thomas Raupach/Woodfin Camp & Associates; P160 E. R. Degginger/Color Pic, Inc.; P161 E. R. Degginger/Color Pic, Inc.; P162 E. R. Degginger/Color Pic, Inc.; P479 Stock Trek/PhotoDisc, Inc.